A COURSE IN

Galois theory

D.J.H. GARLING

*Reader in Mathematical Analysis, University of Cambridge
and Fellow of St John's College, Cambridge*

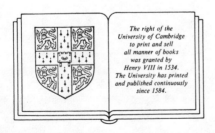

BISHOP MUELLER LIBRARY

Briar Cliff College

SIOUX CITY, IA 51104

The right of the
University of Cambridge
to print and sell
all manner of books
was granted by
Henry VIII in 1534.
The University has printed
and published continuously
since 1584.

CAMBRIDGE UNIVERSITY PRESS

Cambridge

New York   New Rochelle

Melbourne   Sydney

Published by the Press Syndicate of the University of Cambridge
The Pitt Building, Trumpington Street, Cambridge CB2 1RP
32 East 57th Street, New York, NY 10022, USA
10 Stamford Road, Oakleigh, Melbourne 3166, Australia

© Cambridge University Press 1986

First published 1986
Reprinted with corrections 1988

Printed in Great Britain at the University Press, Cambridge

British Library cataloguing in publication data
Garling, D.J.H.
  Galois theory.
  1. Galois theory
  I. Title
  512′.32   QA211

Library of Congress cataloguing in publication data
Garling, D. J. H.
a course in Galois theory.
  1. Galois theory. I. Title.
QA171.G277   1986   512′.32   86-6081

ISBN 0 521 32077 1 hard covers
ISBN 0 521 31249 3 paperback

QA
171
.G277
1986

M P

13332795

CONTENTS

PREFACE

Galois theory is one of the most fascinating and enjoyable branches of algebra. The problems with which it is concerned have a long and distinguished history: the problems of duplicating a cube or trisecting an angle go back to the Greeks, and the problem of solving a cubic, quartic or quintic equation to the Renaissance. Many of the problems that are raised are of a concrete kind (and this, surely, is why it is so enjoyable) and yet the needs of the subject have led to substantial development in many branches of abstract algebra: in particular, in the theory of fields, the theory of groups, the theory of vector spaces and the theory of commutative rings.

In this book, Galois theory is treated as it should be, as a subject in its own right. Nevertheless, in the process, I have tried to show its relationship to various topics in abstract algebra: an understanding of the structures of abstract algebra helps give a shape to Galois theory and conversely Galois theory provides plenty of concrete examples which show the point of abstract theory.

This book comprises two unequal parts. In the first part, details are given of the algebraic background knowledge that it is desirable to have before beginning to study Galois theory. The first chapter is quite condensed: it is intended to jog the memory, to introduce the terminology and notation that is used, and to give one or two examples which will be useful later. In the second chapter, the axiom of choice and Zorn's lemma are described. Algebra is principally concerned with finite discrete operations, and it would have been possible, at the cost of not establishing the existence of algebraic closures, to have avoided all use of the axiom of choice. Mathematicians do, however, need to know about the axiom of choice, and this is an appropriate place to introduce it. A reader who has not met the concepts of this chapter before may omit it (and Chapter 8); preferably, he or she should read through it quite quickly to get some idea of the issues involved, and not worry too much about the details. The third chapter, on rings, is much more important, and should be read rather carefully. It is an important fact that polynomials with integer coefficients and polynomials in several variables enjoy unique factorization, and it is necessary to go beyond principal ideal domains to establish this fact. There are some other special results from abstract algebra that are needed (such as basic properties of soluble groups): most of these are established when the need arises.

The second, more substantial, part is concerned with the theory of fields and with Galois theory, and contains the main material of the book. Of its nature, the theory develops an inexorable momentum. Nevertheless, there are many digressions (for example, concerning irreducibility, geometric constructions, finite fields and the solution of cubic and quartic equations): one of the pleasures of Galois theory is that there are many examples which illustrate and depend upon the general theory, but which also have an interest of their own. The high point of the book is of course the resolution of the problem of when a polynomial is solvable by radicals. I have, however, tried to emphasize (in the final chapter in particular) that this is not the end of the story: the resolution of the problem raises many new problems, and Galois theory is still a lively subject.

Two hundred exercises are scattered through the text. It has been suggested to me that this is rather few: I think that anyone who honestly tries them all will disagree! In my opinion, text-book exercises are often too straightforward, but some of these exercises are quite hard. The successful solution of a challenging problem gives a much better understanding of the powers and limitations of the theory than any number of trivial ones. Remember that mathematics is not a spectator sport!

This book grew out of a course of lectures which I gave for several years at Cambridge University. I have, however, not resisted the temptation to add extra material. A shorter course than the whole book provides can be obtained by omitting Chapter 2, Chapter 8, Section 10.6, Section 18.5 and Chapter 20. I am grateful to all who attended the course, and helped me to improve it. I am particularly grateful to Robert J. H. A. Turnbull, who read and commented helpfully on an early draft and also detected many errors in the final version of this book.

I would like to thank Frank Gerrish for pointing out a large number of corrections to the first printing.

PART 1

Algebraic preliminaries

1

Groups, fields and vector spaces

Galois theory is almost exclusively a branch of algebra; the reader is expected to have some knowledge of algebra, and in particular to have some knowledge of groups and vector spaces. Read through this chapter and make sure that you are familiar with what is in it; if you are not, you should consult standard text-books, such as those by MacDonald[1] on groups and Halmos[2] on vector spaces.

1.1 Groups

Suppose that S is a set. A *law of composition* $\circ$ on S is a mapping from the Cartesian product $S \times S$ into S; that is, for each ordered pair (s_1, s_2) of elements of S there is defined an element $s_1 \circ s_2$ of S.

A *group* G is a non-empty set, with a law of composition $\circ$ on it with the following properties:

    (i) $g_1 \circ (g_2 \circ g_3) = (g_1 \circ g_2) \circ g_3$ for all g_1, g_2 and g_3 in G – that is, composition is *associative*;

   (ii) there is an element e in G (the *unit* or *neutral element*) such that $e \circ g = g \circ e = g$ for each g in G;

  (iii) to each g in G there corresponds an element g^{-1} (the *inverse* of g) such that $g \circ g^{-1} = g^{-1} \circ g = e$.

A group G is said to be *commutative*, or *abelian*, if $g_1 \circ g_2 = g_2 \circ g_1$ for all g_1 and g_2 in G.

The notation that is used for the law of composition $\circ$ varies from situation to situation. Frequently there is no symbol, and elements are simply juxtaposed: $g \circ h = gh$. When G is abelian, it often happens that the

[1] I. D. MacDonald. *The Theory of Groups*, Oxford, 1968.
[2] P. R. Halmos, *Finite-dimensional Vector Spaces*, Springer Verlag, 1974.

law is denoted by $+:g \circ h = g + h$, the neutral element is denoted by 0 and the inverse of an element g is denoted by $-g$.

Let us give some examples of groups. The integers $\mathbb{Z}$ (positive, zero and negative) form a group under addition. This group is abelian, and 0 is the unit element.

If S is a non-empty set, a mapping σ from S into S is called a *permutation* of S if it is one–one (that is, if $\sigma(x) = \sigma(y)$ then $x = y$) and onto (that is, if $y \in S$ there exists x in S such that $\sigma(x) = y$). The set Σ_S of permutations of S is a group under the natural composition of mappings. In detail, $\sigma_1 \circ \sigma_2$ is defined by $\sigma_1 \circ \sigma_2(x) = \sigma_1(\sigma_2(x))$. If $S = \{1, \ldots, n\}$, we write Σ_n for Σ_S. Σ_S is not abelian if S has more than two elements.

A subset H of a group G is a *subgroup* if it is a group under the law of composition defined on G; that is, if h_1 and h_2 are in H, so are $h_1 \circ h_2$ and h_1^{-1}. If G is a group with unit element e, the sets $\{e\}$ and G are subgroups; these are the *trivial* subgroups of G. Here are some examples of subgroups. The sets

$$n\mathbb{Z} = \{nm : m \in \mathbb{Z}\}, \text{ for } n \geqslant 0,$$

are subgroups of $\mathbb{Z}$, and any subgroup of $\mathbb{Z}$ is of this form (why?). The *alternating group* A_n of all permutations in Σ_n which can be written as the product of an even number of transpositions (permutations which interchange two elements, and leave the others fixed) is a subgroup of Σ_n. If s_0 is a fixed element of S, the set

$$\{\sigma \in \Sigma_S : \sigma(s_0) = s_0\}$$

is a subgroup of Σ_S.

Suppose that H is a subgroup of G, and that g is an element of G. We write $H \circ g$ for the set $\{h \circ g : h \in H\}$. Such a set is called a *right coset* of H in G. The collection of right cosets of H is denoted by G/H.

If S is any set, the *order* of S, denoted by $|S|$, is the number of elements of S (a non-negative integer, or $+\infty$). Thus $|\Sigma_n| = n!$. If H is a subgroup of G, and g_1 and g_2 are elements of G, the mapping which sends g to $g \circ g_1^{-1} \circ g_2$ is a permutation of the set G which maps $H \circ g_1$ onto $H \circ g_2$: thus any two right cosets of H in G have the same order. As distinct right cosets are disjoint, $|G| = |G/H| \cdot |H|$ (Lagrange's theorem) and so $|H|$ divides $|G|$. The quantity $|G/H|$ is called the *index* of H in G. Thus A_n has index 2 in Σ_n.

A mapping ϕ from a group G_1 into a group G_2 is a *homomorphism* if $\phi(g_1 \circ g_2) = \phi(g_1) \circ \phi(g_2)$, for all g_1 and g_2 in G. A homomorphism which is one–one is called a *monomorphism*, one which is onto is called an *epimorphism* and one which is both is called an *isomorphism*. If ϕ is a homomorphism of G_1 into G_2, the *image*

$$\phi(G_1) = \{\phi(g) : g \in G_1\}$$

is a subgroup of G_2, and the *kernel*

$$\phi^{-1}(\{e\}) = \{g \in G_1 : \phi(g) = e, \text{ the unit of } G_2\}$$

is a subgroup of G_1. It is, however, a subgroup of a special kind.

A subgroup H of a group G is a *normal* subgroup if $g^{-1} \circ h \circ g \in H$ whenever $g \in G$ and $h \in H$. We write $H \lhd G$ to mean that H is a normal subgroup of G. If $H \lhd G$ and $C_1 = H \circ g_1$ and $C_2 = H \circ g_2$ are right cosets of H in G, then

$$
\begin{aligned}
C_1 \circ C_2 &= \{c_1 \circ c_2 : c_1 \in C_1, c_2 \in C_2\} \\
&= \{h \circ g_1 \circ k \circ g_2 : h, k \in H\} \\
&= \{h \circ (g_1 \circ k \circ g_1^{-1}) \circ g_1 \circ g_2 : h, k \in H\} \\
&= \{h \circ g_1 \circ g_2 : h \in H\} = H \circ (g_1 \circ g_2),
\end{aligned}
$$

so that $C_1 \circ C_2$ is again a right coset of H in G; under this law of composition, G/H is a group (the *quotient group*) with H as unit element, and the natural quotient map $q : G \to G/H$ (which sends g to its right coset $H \circ g$) is an epimorphism with H as kernel.

On the other hand, if ϕ is a homomorphism of G_1 into G_2 with kernel K, then $K \lhd G_1$ and there is an isomorphism $\tilde{\phi}$ from G_1/K onto $\phi(G_1)$ such that $\phi = \tilde{\phi}q$ (the first isomorphism theorem).

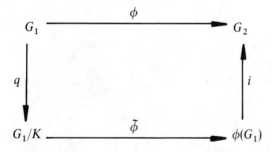

In this diagram, i is the inclusion mapping, which is of course a monomorphism.

Let us give some examples. If G is an abelian group, any subgroup is necessarily normal, and we can form the quotient group. In particular we denote by $\mathbb{Z}_n$ the quotient group $\mathbb{Z}/n\mathbb{Z}$. This is the group of integers (mod n): we identify two integers which differ by a multiple of n. The group $\mathbb{Z}_n$ has order n, for $n > 0$.

There is an epimorphism of Σ_4 onto Σ_3, which can perhaps most easily be described geometrically. Let 1, 2, 3, 4 be four points in a plane, no three of which are collinear (see Fig. 1.1). Denote the line joining i and j by (ij) (there are six such lines), and denote the intersection of (ij) and (kl) by $(ij)(kl)$ (there are three such points of intersection).

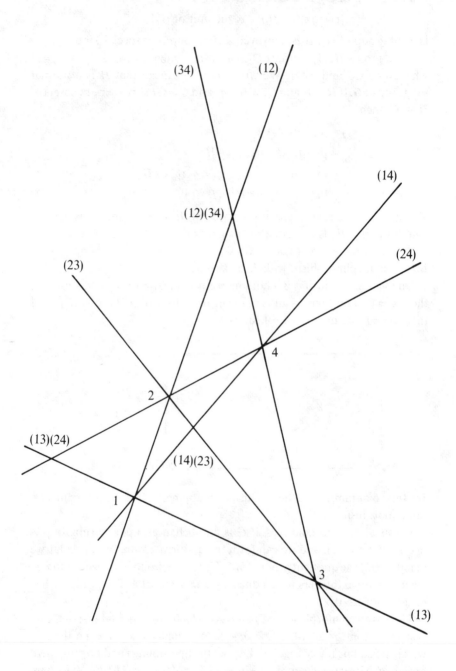

Fig. 1.1

Then any permutation σ of $\{1, 2, 3, 4\}$ defines a permutation of the lines $((ij)$ goes to $(\sigma(i)\sigma(j)))$ and a permutation $\phi(\sigma)$ of the points of intersection $(\phi(\sigma)((ij)(kl)) = (\sigma(i)\sigma(j))(\sigma(k)\sigma(l)))$. It is easy to see that ϕ is a homomorphism, and that (using cycle notation) the kernel of ϕ is

$$\{e, (12)(34), (13)(24), (14)(23)\}.$$

This normal subgroup of Σ_4 is called the Viergruppe, and will be denoted by N. Finally, as

$$|\phi(\Sigma_4)| = |\Sigma_4/N| = |\Sigma_4|/|N| = 6 = |\Sigma_3|,$$

$\phi(\Sigma_4) = \Sigma_3$, and ϕ is an epimorphism.

Suppose that A is a non-empty subset of a group G. We denote by $\langle A \rangle$ the intersection of all those subgroups of G which contain A. $\langle A \rangle$ is a subgroup of G, the smallest subgroup containing A, and we call it the subgroup *generated* by A. If A is a singleton $\{x\}$, we write $\langle x \rangle$ for $\langle A \rangle$; $\langle x \rangle$ is called a *cyclic* subgroup of G. Similarly a group which is generated by a single element is called a *cyclic group*. A cyclic group $\langle x \rangle$ is always abelian. If it is infinite, it is isomorphic to $\mathbb{Z}$; if it has order n, it is isomorphic to $\mathbb{Z}_n$. In this case n is the least positive integer k such that $x^k = e$; n is called the *order* of x. It follows from Lagrange's theorem that if G is finite then the order of x divides $|G|$.

This is the basic group theory that we need; we shall need some more group theory later, and will develop the results as we go along.

Exercises

1.1 Show that an element of a group has exactly one inverse.

1.2 Write out a proof of the first isomorphism theorem.

1.3 Suppose that H is a subgroup of G of index 2. Show that H is a normal subgroup of G.

1.4 Suppose that G has exactly one subgroup H of order k. Show that H is a normal subgroup of G.

1.5 Suppose that H is a normal subgroup of G and that K is a normal subgroup of H. Is K necessarily a normal subgroup of G?

1.6 Show that any permutation of a finite set can be written as a product of transpositions.

1.7 Show that a group G is generated by each of its elements (other than the unit element) if and only if G is a finite cyclic group of prime order.

1.8 Describe the elements of $\mathbb{Z}_n$ which generate $\mathbb{Z}_n$ (for any positive integer n).

1.9 Give an example of a non-abelian group of order 8 all of whose subgroups are normal.

1.10 If $\pi \in \Sigma_n$, let $\varepsilon(\pi) = \prod_{i<j}(\pi(j) - \pi(i))/\prod_{i<j}(j-i)$. Show that ε is a homomorphism of Σ_n onto the multiplicative group with two elements 1 and -1, and that A_n is the kernel of ε.

1.2   Fields

The study of fields is the first main topic of Galois theory. Here we shall do little more than recall the definition.

A field K is a non-empty set with two laws of composition, addition and multiplication, with which the usual arithmetic operations can be carried out. To be precise, first K is an abelian group under addition (written $+$). The neutral or zero element is written as 0 and the additive inverse of x as $-x$. Let K^* denote the set of non-zero elements $K \setminus \{0\}$. Then multiplication satisfies the following.

(a) $0 . x = x . 0 = 0$ for all x in K.

(b) K^* is an abelian group under multiplication. The multiplicative unit is written as 1.

(c) Addition and multiplication are linked by

$$x(y+z) = xy + xz \text{ for all } x, y \text{ and } z \text{ in } K.$$

The fields $\mathbb{Q}$ of *rational* numbers, $\mathbb{R}$ of *real* numbers and $\mathbb{C}$ of *complex* numbers should be familiar and their basic properties will be taken for granted; remember that $\mathbb{R}$ and $\mathbb{C}$ belong as much to analysis as they do to algebra. We shall give one more example of a field here. Recall that the real number $\sqrt{2}$ is not rational. (If it were, we could write $\sqrt{2} = p/q$, where p and q have no common factor. Then $2q^2 = p^2$, so that p would be even, and we could write $p = 2k$. Then $q^2 = 2k^2$, so that q would be even and 2 would be a common factor of p and q, giving a contradiction.) Let K consist of all real numbers of the form $r + s\sqrt{2}$, where r and s are rational. K is clearly an additive subgroup of $\mathbb{R}$. Also

$$(r_1 + s_1\sqrt{2})(r_2 + s_2\sqrt{2}) = (r_1 r_2 + 2s_1 s_2) + (r_1 s_2 + r_2 s_1)\sqrt{2},$$

and if $r + s\sqrt{2} \neq 0$ then $r - s\sqrt{2} \neq 0$, so that $r^2 - 2s^2 \neq 0$. Thus if we set $u = r/(r^2 - 2s^2)$, $v = -s/(r^2 - 2s^2)$,

$$(r + s\sqrt{2})(u + v\sqrt{2}) = 1.$$

Consequently K^* is an abelian group under multiplication, and so K is a *subfield* of $\mathbb{R}$.

This field K is a very simple example of the sort of field that we shall consider later. Notice that the only difficulty involved finding a multiplicative inverse. Why did we consider $r - s\sqrt{2}$, and where did it come from?

Exercises

1.11 Every field has at least two elements. Show that there is a field with exactly two elements.

1.12 Let $\mathbb{Z}_n$ be equipped with a multiplication by defining $(a + n\mathbb{Z}) \circ (b + n\mathbb{Z}) = ab + n\mathbb{Z}$. Show that this is well defined, and that with this multiplication $\mathbb{Z}_n$ is a field if and only if n is a prime number.

1.13 Which of the following subsets of $\mathbb{C}$ are subfields of $\mathbb{C}$?

    (i) $\{a + ib : a, b \in \mathbb{Q}\}$.

    (ii) $\{a + \omega b : a, b \in \mathbb{Q}, \omega = \frac{1}{2}(-1 + \sqrt{3}\,i)\}$.

    (iii) $\{a + 2^{1/3}b : a, b \in \mathbb{Q}\}$.

    (iv) $\{a + 2^{1/3}b + 4^{1/3}c : a, b, c \in \mathbb{Q}\}$.

1.3   Vector spaces

We are now in a position to define the notion of a vector space. Suppose that K is a field. A set V is a *vector space over K* if first it is an abelian group under addition and secondly there is a mapping $(\alpha, x) \to \alpha x$ from $K \times V$ into V which satisfies

    (a) $\alpha(x + y) = \alpha x + \alpha y$,
    (b) $(\alpha + \beta)x = \alpha x + \beta x$,
    (c) $(\alpha\beta)x = \alpha(\beta x)$, and
    (d) $1 . x = x$

for all α, β in K and x, y in V.

As an example, let S be a non-empty set, and let K^S denote the set of all mappings from S into K. If f and g are in K^S, define $f + g$ by

$$(f + g)(s) = f(s) + g(s), \text{ for } s \text{ in } S,$$

and, if $\alpha \in K$, define αf by

$$(\alpha f)(s) = \alpha(f(s)), \text{ for } s \text{ in } S.$$

Then it is easy to verify that the axioms are satisfied. If $S = \{1, \ldots, n\}$, we write K^n for K^S and, if $x \in K^n$, write

$$x = (x_1, \ldots, x_n),$$

where x_j is the value of x at j.

In fact, we need surprisingly little of the theory of vector spaces. The key is the idea of *dimension*; as we shall see, this turns out to be remarkably powerful. Suppose that V is a vector space over K. A subset W of V is a *linear subspace* if it is a vector space under the operations defined on V; for this, it is sufficient that if x and y are in W and α is in K then $x + y \in W$ and $\alpha x \in W$. If A is a non-empty subset of V, the *span* of A, denoted by span (A), is the intersection of the linear subspaces containing A; it is a linear subspace of V, and is the smallest linear subspace containing A. If span $(A) = V$, we say that A *spans* V.

We now turn to linear dependence and linear independence. A subset A of a vector space V over K is *linearly dependent over K* if there are finitely many distinct elements $x_1, \ldots, x_k$ of A and elements $\lambda_1, \ldots, \lambda_k$ of K, *not all zero*, such that

$$\lambda_1 x_1 + \cdots + \lambda_k x_k = 0;$$

if A is not linearly dependent over K, A is *linearly independent over K*. Note that, even if A is infinite, the sums which we consider are finite. If A is finite and $A = \{x_1, \ldots, x_n\}$, *where the x_i are distinct*, A is linearly independent over K if it follows from

$$\lambda_1 x_1 + \cdots + \lambda_n x_n = 0$$

that $\lambda_1 = \lambda_2 = \cdots = \lambda_n = 0$.

A subset A of a vector space V over K is a *basis* for V if it is linearly independent and spans V. We shall see in the next chapter that every vector space has a basis. In fact, our main interest is in finite-dimensional vector spaces; let us consider them now.

A vector space V over K is *finite dimensional* if there exists a finite subset of V which spans V. First we show that a finite-dimensional space has a finite basis; this is a consequence of the following theorem:

Theorem 1.1 *Suppose that A is a finite subset of a vector space V over K which spans V, and that C is a linearly independent subset of A (C may be empty). There exists a basis B of V with $C \subseteq B \subseteq A$.*

Proof. Consider the collection J of all subsets of A which contain C and are linearly independent. Since $|A| < \infty$, there exists a B in J with a maximum number of elements. B is independent and $C \subseteq B \subseteq A$; it remains to show that B spans V.

Let $B = \{b_1, \ldots, b_n\}$, where the b_i are distinct. If $a \in A \setminus B$, $B \cup \{a\}$ is linearly dependent (by the maximality of $|B|$) and so there exist $\lambda_0, \ldots, \lambda_n$ in K, not all zero, such that

$$\lambda_0 a + \lambda_1 b_1 + \cdots + \lambda_n b_n = 0.$$

Further, $\lambda_0 \neq 0$, for otherwise $b_1, \ldots, b_n$ would be linearly dependent. Thus

$$a = -\lambda_0^{-1} \lambda_1 b_1 - \lambda_0^{-1} \lambda_2 b_2 - \cdots - \lambda_0^{-1} \lambda_n b_n$$

and $a \in \text{span}(B)$. Consequently $A \subseteq \text{span}(B)$, and so $\text{span}(A) \subseteq \text{span}(B)$. As $\text{span}(A) = V$, the theorem is proved.

We would now like to define the dimension of a finite-dimensional vector space as the number of elements in a basis. In order to do this, we need to show that any two bases have the same number of elements. This follows from the next theorem.

Theorem 1.2 *Suppose that V is a vector space over K. If A spans V and C is a linearly independent subset of V, then $|C| \leqslant |A|$.*
Proof. The result is trivially true if $|A| = \infty$, and so we may suppose that $|A| < \infty$. If $|C| = \infty$, there is a finite subset D of C with $|D| > |A|$. As D is again linearly independent, it is sufficient to prove the result when $|C| < \infty$. Theorem 1.2 is therefore a consequence of the following:

Theorem 1.3 (The Steinitz exchange theorem) *Suppose that $C = \{c_1, \ldots, c_r\}$ is a linearly independent subset (with r distinct elements) of a vector space V over K, and that $A = \{a_1, \ldots, a_s\}$ is a set (with s distinct elements) which spans V. Then there exists a set D, with $C \subseteq D \subseteq A \cup C$, such that $|D| = s$ and D spans V.*
Proof. We prove this by induction on r. The result is trivially true for $r = 0$ (take $D = A$). Suppose that it is true for $r - 1$. As the set $C_0 = \{c_1, \ldots, c_{r-1}\}$ is linearly independent, there exists a set D_0 with $C_0 \subseteq D_0 \subseteq A \cup C_0$ such that $|D_0| = s$, and D_0 spans V. By relabelling A if necessary, we can suppose that

$$D_0 = \{c_1, \ldots, c_{r-1}, a_r, a_{r+1}, \ldots, a_s\}.$$

If s were equal to $r - 1$, we would have $D_0 = C_0$; but $c_r \in \text{span}(D_0)$, so that we could write

$$c_r = \sum_{i=1}^{r-1} \gamma_i c_i,$$

contradicting the linear independence of C. Thus $s \geqslant r$. As $c_r \in \text{span}(D_0)$, we can write

$$c_r = \sum_{i=1}^{r-1} \gamma_i c_i + \sum_{j=r}^{s} \alpha_j a_j.$$

Not all α_j can be zero, for again this would contradict the linear independence of C. By relabelling if necessary, we can suppose that $\alpha_r \neq 0$. Let $D = \{c_1, \ldots, c_r, a_{r+1}, \ldots, a_s\}$. Then

$$a_r = \alpha_r^{-1} \left(c_r - \sum_{i=1}^{r-1} \gamma_i c_i - \sum_{j=r+1}^{s} \alpha_j a_j \right)$$

so that $a_r \in \text{span}(D)$. Thus
$$\text{span}(D) \supseteq \{c_1, \ldots, c_{r-1}, a_r, a_{r+1}, \ldots, a_s\} = D_0$$
and so $\text{span}(D) \supseteq \text{span}(D_0) = V$.

This completes the proof.

Corollary (to Theorem 1.2) *Any two bases of a finite-dimensional vector space have the same finite number of elements.*

We now define the *dimension* of a finite-dimensional vector space V over K to be the number of elements in a basis. We denote the dimension of V by dim V. Here is one simple but important result:

Theorem 1.4 *Suppose that U is a linear subspace of a finite-dimensional vector space V over K. Then* dim $U \leqslant$ dim V, *and* dim $U =$ dim V *if and only if $U = V$.*

Proof. Let A be a basis for U, and let C be a finite set which spans V. Considered as a subset of V, A is linearly independent, and so by Theorem 1.1 there is a basis B of V with $A \subseteq B \subseteq A \cup C$. Thus
$$\text{dim } U = |A| \leqslant |B| = \text{dim } V.$$
If dim $U =$ dim V, we must have $A = B$, so that A spans V and $U = V$; of course if $U = V$, dim $U =$ dim V.

Corollary 1 *Suppose that A is a finite subset of a finite-dimensional vector space V over K. If $|A| >$ dim V, A is linearly dependent.*

Proof. Let $U = \text{span}(A)$. If A were linearly independent, A would be a basis for U, so that dim $U = |A|$. But dim $U \leqslant$ dim V, giving a contradiction.

Suppose that V_1 and V_2 are vector spaces over the same field K. A mapping ϕ from V_1 into V_2 is called a *linear mapping* if
$$\phi(x + y) = \phi(x) + \phi(y),$$
$$\phi(\lambda x) = \lambda \phi(x)$$
for all x and y in V_1 and all λ in K. The study of linear mappings is an essential part of the study of vector spaces; for our purposes we shall only need one further corollary to Theorem 1.4.

Corollary 2 *Suppose that V_1 and V_2 are vector spaces over K and that ϕ is a linear mapping of V_1 into V_2. If* dim $V_1 >$ dim V_2, *ϕ is not one–one, and there exists a non-zero x in V_1 such that $\phi(x) = 0$.*

Proof. Let $n =$ dim V_2. As dim $V_1 >$ dim V_2, there exist $n + 1$ linearly independent vectors $x_1, \ldots, x_{n+1}$ in V_1. Then, by Corollary 1, $\{\phi(x_1), \ldots, \phi(x_{n+1})\}$ is linearly dependent in V_2, and so there exist $\lambda_1, \ldots, \lambda_{n+1}$ in K, not all zero, such that

$$\lambda_1 \phi(x_1) + \cdots + \lambda_{n+1} \phi(x_{n+1}) = 0$$

But

$$\lambda_1 \phi(x_1) + \cdots + \lambda_{n+1} \phi(x_{n+1}) = \phi(\lambda_1 x_1 + \cdots + \lambda_{n+1} x_{n+1}),$$

since ϕ is linear, and

$$x = \lambda_1 x_1 + \cdots + \lambda_{n+1} x_{n+1} \neq 0$$

since $\{x_1, \ldots, x_{n+1}\}$ is linearly independent. As $\phi(x) = 0 = \phi(0)$, ϕ is not one–one.

Exercises

1.14 In K^n, let $e_j = (0, \ldots, 0, 1, 0, \ldots, 0)$, where the 1 occurs in the jth position. Let $f_j = e_1 + \cdots + e_j$.
   (a) Show that $\{e_1, \ldots, e_n\}$ is a basis for K^n.
   (b) Show that $\{f_1, \ldots, f_n\}$ is a basis for K^n.
   (c) Is $\{e_1, f_1, f_2, \ldots, f_n\}$ a basis for K^n?

1.15 Suppose that S is infinite. For each s in S, let $e_s(t) = 1$ if $s = t$, and let $e_s(t) = 0$ otherwise. Is $\{e_s : s \in S\}$ a basis for K^S?

1.16 $\mathbb{R}$ can be considered as a vector space over $\mathbb{Q}$. Show that $\mathbb{R}$ is *not* finite dimensional over $\mathbb{Q}$. Can you find an infinite subset of $\mathbb{R}$ which is linearly independent over $\mathbb{Q}$?

1.17 Suppose that K is an infinite field and that V is a vector space over K. Show that it is not possible to write $V = \bigcup_{i=1}^n U_i$, where $U_1, \ldots, U_n$ are proper linear subspaces of V.

2

The axiom of choice, and Zorn's lemma

The study of algebra is very largely concerned with considering finitely many operations on finitely many objects; even when induction is used, as in Theorem 1.3, at any one time we consider only finitely many elements. There are, however, one or two situations when we need to consider infinitely many objects simultaneously; in order to be able to do this, we have to appeal to the axiom of choice.

2.1 The axiom of choice

In its simplest form, the axiom of choice can be expressed as follows. Suppose that $\{E_\alpha\}_{\alpha \in A}$ is an indexed family of sets, and that each of the sets E_α is not empty. Then the axiom of choice says that the Cartesian product $\prod_{\alpha \in A}(E_\alpha)$ is also non-empty: that is, there exists an element $(c_\alpha)_{\alpha \in A}$ in the product. In these terms, the axiom of choice may seem rather self-evident: each E_α is not empty, and so we can find a suitable c_α. The point is that we want to be able to make this choice simultaneously. The more one thinks about it, the more one discovers that this is a rather strong requirement. The axiom of choice is a genuine axiom of set theory; most mathematicians accept it and use it, as we certainly shall, but there are those who do not. Arguments which use the axiom of choice, or one of its equivalents, have a character of their own. You should avoid using it unless it is really necessary.

2.2 Zorn's lemma

In the form in which we have described it, the axiom of choice is a rather unwieldly tool. There are many statements which are equivalent to the axiom of choice (in the sense that they can be deduced using the axiom of choice, and the axiom of choice can be deduced from them). Thus an equivalent statement is that every set can be 'well ordered' (see Exercise 2.2

for a definition): this is fundamental to the theory of ordinals, and leads to the idea of 'transfinite induction'. Some fifty years ago, this was the most popular and effective way of using the axiom of choice.

More recently, it has become customary to use another equivalent of the axiom of choice, namely Zorn's lemma. This is a technical result concerning partially ordered sets, which proves to be rather simple to use in practice. In order to state it, we need to say something about partially ordered sets.

A relation $\leqslant$ on a set S is said to be a *partial order* if

    (a) $x \leqslant x$ for all x in S,
    (b) if $x \leqslant y$ and $y \leqslant z$ then $x \leqslant z$, and
    (c) if $x \leqslant y$ and $y \leqslant x$ then $x = y$.

For example, if S is a collection of subsets of a set X, the relation $E \leqslant F$ if $E \subseteq F$ is a partial order on S ('ordering by inclusion').

A partially ordered set S is *totally ordered* if any two elements can be compared: if x and y are in S then either $x \leqslant y$ or $y \leqslant x$. A non-empty subset C of a partially ordered set S is a *chain* if it is totally ordered in the ordering inherited from S.

If A is a subset of a partially ordered set S, an element x of S is an *upper bound* for A if $a \leqslant x$ for each a in A. An upper bound may or may not belong to A: A may have many upper bounds, or none at all. For example, let S be the collection of finite subsets of an infinite set X, ordered by inclusion. S itself has no upper bound, and a subset of S has an upper bound in S if and only if it is finite.

Finally, an element x of a partially ordered set S is *maximal* in S if, whenever $x \leqslant y$, we must have $x = y$. In other words x is maximal if there are no larger elements. A maximal element need not be an upper bound for S; there may be other elements which cannot be compared with x. S may well have many maximal elements.

We are now in a position to state Zorn's lemma.

Zorn's lemma *Suppose that S is a partially ordered set with the property that every chain in S has an upper bound. Then S has at least one maximal element.*

We shall not show how to deduce this from the axiom of choice. A proof can be found in Halmos[1]. After working through the proof of Theorem 2.1, you should be able to tackle exercises 2.1 and 2.2.

2.3   The existence of a basis

In our study of Galois theory, we shall only need to apply Zorn's lemma three times (in Theorems 8.2 (via Theorem 3.14), 8.3 and 18.5). To

---

[1] P. R. Halmos, *Naive Set Theory*, Springer-Verlag, 1974.

illustrate how Zorn's lemma can be used, let us show that every vector space has a basis.

Theorem 2.1 *Suppose that A is a subset of a vector space V over K which spans V and that C is a linearly independent subset of A (C may be empty). There exists a basis B of V with $C \subseteq B \subseteq A$.*

We have proved this in the case that A is finite in Theorem 1.1 (and made essential use of the finiteness of A). Taking $A = V$ and C the empty set, we see that this theorem implies that every vector space has a basis.

Proof. Let S denote the collection of subsets of A which are linearly independent and contain C. Order S by inclusion. Suppose that T is a chain in S. Let $E = \bigcup_{D \in T} D$. E is a subset of A which contains C. Suppose that $x_1, \ldots, x_n$ are distinct elements of E. From the definition of E, there are sets $D_1, \ldots, D_n$ in T such that $x_i \in D_i$ for $1 \leqslant i \leqslant n$. Since T is a chain, there exists j, with $1 \leqslant j \leqslant n$, such that $D_i \subseteq D_j$ for $1 \leqslant i \leqslant n$. Consequently $x_1, \ldots, x_n$ are all in D_j. As D_j is linearly independent, $\{x_1, \ldots, x_n\}$ is linearly independent. As this holds for any finite subset of E, E is linearly independent. Thus $E \in S$. E is clearly an upper bound for T, and so every chain in S has an upper bound.

We can therefore apply Zorn's lemma, and conclude that S has a maximal element B. B is linearly independent and $C \subseteq B \subseteq A$; it remains to show that span $(B) = V$. Since span $(A) = V$, it is enough to show that $A \subseteq$ span (B). If not, there exists a in A which does not belong to span (B). Let $B_0 = \{a\} \cup B$. We shall show that B_0 is linearly independent. Suppose that

$$\lambda_0 a + \lambda_1 b_1 + \cdots + \lambda_n b_n = 0$$

where $b_1, \ldots, b_n$ are distinct elements of B. If $\lambda_0 \neq 0$,

$$a = -\lambda_0^{-1}(\lambda_1 b_1 + \cdots + \lambda_n b_n)$$

contradicting the fact that $a \notin$ span (B). Thus

$$\lambda_1 b_1 + \cdots + \lambda_n b_n = 0.$$

As B is linearly independent, $\lambda_1 = \cdots = \lambda_n = 0$. Thus B_0 is linearly independent. Consequently $B_0 \in S$. But $B_0 \supseteq B$, and $B_0 \neq B$, contradicting the maximality of B. This completes the proof.

The proof of this theorem should be compared with the proof of Theorem 1.1. In fact, the proofs are very similar: the counting argument of Theorem 1.1 is replaced by a maximality argument.

Exercises

2.1 Show that the axiom of choice is a consequence of Zorn's lemma. (Hint: Suppose that $\{E_\alpha\}_{\alpha \in A}$ is a family of non-empty sets. Take S to be all *pairs* $(B, (c_\beta)_{\beta \in B})$ where B is a subset of A, and $c_\beta \in E_\beta$ for $\beta \in B$. Partially order S by setting $(B, (c_\beta)_{\beta \in B}) \leqslant (C, (c'_\gamma)_{\gamma \in C})$ if $B \subseteq C$ and $c_\beta = c'_\beta$ for $\beta \in B$. Now carry through the procedures of Theorem 2.1.)

2.2 A total order on a set S is said to be a 'well-ordering' if every non-empty subset of S has a least element. Use Zorn's lemma to show that every non-empty set can be given a well-ordering. (Hint: Define a partial order $\prec$ on all pairs $(T, \leqslant)$, where T is a subset of S, and $\leqslant$ is a well-ordering on T, by saying that

$$(T_1, \leqslant_1) \prec (T_2, \leqslant_2)$$

if first $T_1 \subseteq T_2$, secondly the orderings $\leqslant_1$ and $\leqslant_2$ coincide on T_1 and thirdly

$$P_t = \{x \in T_2 : x \leqslant_2 t\}$$

is contained in T_1 whenever t is in T_1. Apply Zorn's lemma to this.)

2.3 Suppose that $(A, \leqslant)$ is an infinite well-ordered set. Show that there is a unique element a such that $\{x : x < a\}$ is infinite, while $\{x : x < b\}$ is finite for each $b < a$. Suppose that A is uncountable. Show that there is a unique element c such that $\{x : x < c\}$ is uncountable, while $\{x : x < d\}$ is countable for each $d < c$.

2.4 Suppose that $(A, \leqslant)$ and $(B, \leqslant)$ are two well-ordered sets. Show that one (and only one) of the following must occur:

(i) there is a unique order-preserving bijection $i : A \to B$;
(ii) there exists a unique element a in A and a unique order-preserving bijection $i : \{x : x < a\} \to B$;
(iii) there exists a unique element b in B and a unique order-preserving bijection $i : A \to \{y : y < b\}$.

3

Rings

The second main topic of Galois theory is the study of polynomials. The collection of all polynomials with coefficients in a given field forms a ring, and rings provide a good setting for the study of factorization and divisibility. This chapter is concerned with developing the properties of rings that we shall need; the material is to some extent of a preliminary nature, and you may well be familiar with much of it.

3.1   Rings

A *commutative ring with a 1* is a non-empty set R with two laws of composition: addition and multiplication. Under addition, R is an abelian group, with neutral element 0. As far as multiplication is concerned, the following conditions must be satisfied:

(a) $(rs)t = r(st)$,
(b) $rs = sr$,
(c) $(r+s)t = rt + st$,
(d) there exists an *identity element* 1, different from 0, such that $r \cdot 1 = r$, for all r, s and t in R.

Algebraists study rings which are not commutative under multiplication (for example, rings of matrices) and rings which do not possess an identity element 1. Such rings will not concern us: all our rings are commutative, and possess an identity element 1. For this reason, we shall abbreviate 'commutative ring with a 1' to *ring*.

Let us give some examples.

1. A field is a ring.

2. The integers $\mathbb{Z}$ form a ring under the usual operations of addition and multiplication.

3. The set $\mathbb{Z} + i\mathbb{Z}$ of all complex numbers of the form $m + in$, with m and n integers, forms a ring, under the usual operations.

4. The set $\mathbb{Z} + i\sqrt{5}\,\mathbb{Z}$ of all complex numbers of the form $m + i\sqrt{5}\,n$, with m and n integers, forms a ring.

5. Suppose that R is a ring. Let $R[x]$ denote all sequences

$$(a_0, a_1, \ldots, a_n, 0, 0, 0, \ldots)$$

of elements of R which are zero from some point on. We define addition coordinate by coordinate and define the product of

$$a = (a_0, a_1, \ldots, a_n, 0, 0, 0, \ldots)$$

and

$$b = (b_0, b_1, \ldots, b_m, 0, 0, 0, \ldots)$$

to be

$$ab = (a_0 b_0, a_1 b_0 + a_0 b_1, a_2 b_0 + a_1 b_1 + a_0 b_2, \ldots, 0, 0, \ldots).$$

It is straightforward, but tedious, to verify that the conditions for being a ring are satisfied, with zero element

$$\mathbf{0} = (0, 0, \ldots)$$

and unit element

$$\mathbf{1} = (1, 0, 0, \ldots).$$

Now let x denote the element

$$x = (0, 1, 0, 0, \ldots).$$

It is readily verified from the definition of multiplication that

$$x^r = (0, 0, \ldots, 0, 1, 0, \ldots)$$

where the single 1 occurs in the $(r + 1)$th place. Thus if

$$a = (a_0, a_1, \ldots, a_n, 0, 0, \ldots)$$

we have

$$a = a_0 + a_1 x + \cdots + a_n x^n,$$

where $a_j = (a_j, 0, 0, \ldots)$. Thus $R[x]$ represents all polynomial expressions in one variable or indeterminate x. We can consider the map $a \to a$ as embedding R as a subring of $R[x]$. We shall identify R with its image, and write

$$a = a_0 + a_1 x + \cdots + a_n x^n.$$

If $a \neq 0$, we can write a in this form, with $a_n \neq 0$; n is then called the *degree* of a.

6. We can also consider polynomials in more than one variable. In the case of finitely many variables, we can proceed as for $R[x]$ by considering arrays of elements of R (with only finitely many non-zero terms), or alternatively can proceed inductively and define

$$R[x_1, \ldots, x_n] = (R[x_1, \ldots, x_{n-1}])[x_n].$$

If S is an infinite set, we define the ring $R[X_S]$ to be the union of all polynomial rings $R[x_{s_1}, \ldots, x_{s_n}]$, with $s_1, \ldots, s_n$ in S, with the obvious laws of composition.

Exercise

3.1 Suppose that S is a set and R is a ring. Let R^S denote the set of all mappings from S to R. Show that R^S is a ring, under the operations defined by

$$(f+g)(s) = f(s) + g(s), \quad (fg)(s) = f(s)g(s).$$

Show that if S has more than one element then there exist non-zero elements f and g in R^S for which $fg = 0$.

3.2   Integral domains

The special properties that a ring may have are many and various: in this section we meet the first of them.

A ring is said to be an *integral domain* if whenever $rs = 0$ it follows that one of r and s must be zero. The first four examples of the previous section are integral domains: $R[x]$ is an integral domain if R is (if $p = a_0 + a_1 x + \cdots + a_n x^n$ and $q = b_0 + b_1 x + \cdots + b_m x^m$, with $a_n \neq 0$ and $b_m \neq 0$, the coefficient of x^{m+n} in pq is $a_n b_m$, which is non-zero).

Starting from the integers $\mathbb{Z}$, we can construct the field of rational numbers $\mathbb{Q}$. An exactly similar procedure can be carried out for any integral domain, as we shall now describe.

Let R be an integral domain, and let R^* denote the non-zero elements of R. Intuitively, a fraction is an expression of the form r/s, where $r \in R$ and $s \in R^*$. But different expressions can represent the same fraction: $r_1/s_1 = r_2/s_2$ if $r_1 s_2 = r_2 s_1$. It is therefore necessary to identify two expressions if they represent the same fraction. This leads us to proceed as follows. On $R \times R^*$ we define a relation by setting $(r_1, s_1) \sim (r_2, s_2)$ if $r_1 s_2 = r_2 s_1$. Clearly

(a) $(r, s) \sim (r, s)$, and

(b) $(r_1, s_1) \sim (r_2, s_2)$ if and only if $(r_2, s_2) \sim (r_1, s_1)$. Further

(c) if $(r_1, s_1) \sim (r_2, s_2)$ and $(r_2, s_2) \sim (r_3, s_3)$ then $(r_1, s_1) \sim (r_3, s_3)$.
     For $r_1 s_2 = r_2 s_1$ and $r_2 s_3 = r_3 s_2$, so that $r_1 s_2 s_3 = r_2 s_3 s_1 = r_3 s_2 s_1$, and so $(r_1 s_3 - r_3 s_1)s_2 = 0$. Since $s_2 \neq 0$ and R is an integral domain, $r_1 s_3 = r_3 s_1$.

Thus $\sim$ is an *equivalence relation* on $R \times R^*$. Recall that a subset E of $R \times R^*$ is an *equivalence class* if E is non-empty and whenever $x \in E$ then $E = \{y : x \sim y\}$, and that it follows from (a), (b) and (c) that distinct equivalence classes are disjoint and that the union of all the equivalence classes is $R \times R^*$. Let F denote the collection of equivalence classes. If $(r, s) \in R \times R^*$, let r/s denote the equivalence class to which (r, s) belongs. We now define algebraic operations on F in an obvious way:

$$r_1/s_1 + r_2/s_2 = (r_1 s_2 + r_2 s_1)/(s_1 s_2),$$
$$(r_1/s_1)(r_2/s_2) = (r_1 r_2)/(s_1 s_2)$$

It is straightforward to verify that these do not depend upon the choice of representatives, that they make sense (as $s_1 s_2 \neq 0$), and that under these operations F becomes an integral domain. If $r/s \neq 0$, $r \neq 0$, and s/r is the multiplicative inverse of r/s: thus F is a field. Finally the map $r \to r/1$ embeds R as a *subring* of F. F is called the *field of fractions* of R.

Suppose now that K is a field. Then the polynomial ring $K[x_1, \ldots, x_n]$ is an integral domain. We denote the corresponding field of fractions by $K(x_1, \ldots, x_n)$: the elements of this field are called *rational expressions in $x_1, \ldots, x_n$ over K*. Similarly we denote the field of fractions of $K[X_S]$ by $K(X_S)$.

Exercises

3.2 Suppose that R is an integral domain, with field of fractions F. Show that the field of fractions of $R[x_1, \ldots, x_n]$ can be identified naturally with $F(x_1, \ldots, x_n)$.

3.3 Show that an integral domain with a finite number of elements is always a field.

3.3   Ideals

Suppose that R and S are two rings. A mapping ϕ from R to S is called a *ring homomorphism* if

(a) $\phi(r_1 + r_2) = \phi(r_1) + \phi(r_2)$,
(b) $\phi(r_1 r_2) = \phi(r_1)\phi(r_2)$, and
(c) $\phi(1_R) = 1_S$

for r_1, r_2 in R.

A homomorphism which is one–one is called a *monomorphism*, one which is onto is called an *epimorphism* and one which is both is called an *isomorphism*.

The image $\phi(R)$ is a subring of S. The kernel $\phi^{-1}(\{0\})$ is not (since $1_R \notin \phi^{-1}(\{0\})$), but has rather different properties. A non-empty subset J of a ring R is said to be an *ideal* if the following conditions hold:

(i) if r and s are in J, so is $r + s$;
(ii) if $r \in R$ and $s \in J$, then $rs \in J$.

Note that if $s \in J$, then $-s = (-1)s \in J$, so that J is a subgroup of the additive group $(R, +)$. The ring R is an ideal in R; all ideals other than R are called *proper ideals*.

Let us consider some examples.

1. The sets $n\mathbb{Z}$ are ideals in the ring $\mathbb{Z}$, and any ideal of $\mathbb{Z}$ is of this form.

2. Suppose that A is a non-empty subset of a ring R. We denote by (A) the intersection of all ideals which contain A. (A) is called the *ideal generated by* A. Further,

$$(A) = \{r \in R : r = r_1 a_1 + \cdots + r_n a_n; r_i \in R, a_i \in A\},$$

for every element in the set on the right-hand side must be in (A), and it is easy to see that this set is an ideal which contains A.

3. We write $(a_1, \ldots, a_n)$ for $(\{a_1, \ldots, a_n\})$. An ideal (a) generated by a single element is called a *principal* ideal. (a) consists of all multiples of a by elements of R.

If ϕ is a ring homomorphism from R into S, and $\phi(r) = \phi(s) = 0$, then

$$\phi(r+s) = \phi(r) + \phi(s) = 0.$$

Also if $t \in R$, $\phi(tr) = \phi(t)\phi(r) = \phi(t)0 = 0$; thus the kernel is an ideal. As $\phi(1_R) \neq 0$, the kernel is a proper ideal.

If J is a proper ideal in R, J is a normal subgroup of $(R, +)$; we can construct the quotient group R/J. We can also define the product of two (right) cosets: if C_1 and C_2 are two cosets, we define

$$C_1 C_2 = \{c_1 c_2 + j : c_1 \in C_1, c_2 \in C_2, j \in J\}.$$

If $c_1' c_2' + j'$ and $c_1 c_2 + j$ are elements of $C_1 C_2$,

$$(c_1' c_2' + j') - (c_1 c_2 + j) = c_1'(c_2' - c_2) + c_2(c_1' - c_1) + (j' - j) \in J,$$

so that $C_1 C_2$ is a coset. It is straightforward to verify that with these operations R/J is a ring, with unit $J + 1_R$, and that the quotient map $q : R \to R/J$ is a ring homomorphism, with kernel J. As an example, the quotient $\mathbb{Z}_n = \mathbb{Z}/n\mathbb{Z}$ is a ring, the *ring of integers* (mod n). Just as for groups, we have an isomorphism theorem:

Theorem 3.1. *Suppose that ϕ is a ring homomorphism from a ring R to a ring S, with kernel J. There is a ring isomorphism $\tilde{\phi}$ from R/J onto $\phi(R)$ such that $\phi = \tilde{\phi} \circ q$.*

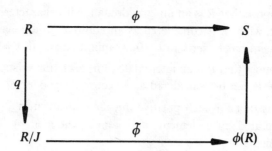

Proof. ϕ is a group homomorphism from $(R, +)$ to $(S, +)$, so that by the first isomorphism theorem for groups there is a group isomorphism $\tilde{\phi}: R/J \to \phi(R)$ such that $\phi = \tilde{\phi}q$. If C_1 and C_2 are two cosets of J, we can write $C_1 = q(x_1) = J + x_1$, $C_2 = q(x_2) = J + x_2$ for some x_1 and x_2 in R. Then $C_1 C_2 = q(x_1 x_2) = J + x_1 x_2$, and so

$$\tilde{\phi}(C_1 C_2) = \tilde{\phi}(q(x_1 x_2)) = \phi(x_1 x_2) = \phi(x_1)\phi(x_2)$$
$$= \tilde{\phi}(q(x_1))\tilde{\phi}(q(x_2)) = \tilde{\phi}(C_1)\tilde{\phi}(C_2).$$

Similarly

$$\tilde{\phi}(J + 1_R) = \tilde{\phi}q(1_R) = \phi(1_R) = 1_S.$$

Let us give an important example of a ring homomorphism. Suppose that R is a ring, with unit element 1_R, and zero element 0_R. We define a map ϕ from $\mathbb{Z}$ into R. Let $\phi(0) = 0_R$, let $\phi(1) = 1_R$, and if n is a positive integer let $\phi(n) = 1_R + \cdots + 1_R$, the sum being taken over n terms. Clearly $\phi(n + m) = \phi(n) + \phi(m)$. If n is a negative integer, let $\phi(n) = -\phi(-n)$. It is then straightforward to show that $\phi(n + m) = \phi(n) + \phi(m)$ for all n and m, and, using the distributive law for rings, that $\phi(mn) = \phi(m)\phi(n)$. Thus ϕ is a ring homomorphism from $\mathbb{Z}$ into R.

There are now two possibilities. Either ϕ is one–one, in which case $\phi(\mathbb{Z})$ is isomorphic to $\mathbb{Z}$, or ϕ fails to be one–one, in which case, by Theorem 3.1, $\phi(\mathbb{Z})$ is isomorphic to the finite ring $\mathbb{Z}_n$, for some n. In the former case, we say that R has *characteristic* 0, in the latter that R has *characteristic* n. We write char R for the characteristic of R.

Suppose that R has non-zero characteristic n, and that n is not a prime number. We can write $n = pq$, where $1 < p < n$. Then $\phi(p) \neq 0$ and $\phi(q) \neq 0$ (since p and q do not belong to $n\mathbb{Z}$) but $\phi(p)\phi(q) = \phi(pq) = \phi(n) = 0$. Thus if R is an integral domain, its characteristic must either be 0 or a prime number.

Exercises

3.4 Suppose that a and b are elements of a ring R for which $(a, b) = R$. Show that $(a^m, b^n) = R$ for any positive integers m and n.

3.5 Suppose that R is an integral domain with characteristic k. Show that, when R is considered as an additive group, every non-zero element has order k (if $k>0$) or infinite order (if $k=0$).

3.6 Suppose that R is an integral domain of characteristic $k>0$. Show how R can be considered as a vector space over $\mathbb{Z}_k$.

3.7 Construct for each positive integer n an ideal in $\mathbb{Z}[x]$ which is generated by n elements and is not generated by fewer than n elements.

3.8 Suppose that K is a field. If $f=a_0+a_1x+\cdots+a_nx^n \in K[x]$ and $k \in K$, let $(\phi(f))(k)=a_0+a_1k+\cdots+a_nk^n$. Show that ϕ is a ring homomorphism from $K[x]$ to K^K. Show that if K is finite then ϕ is an epimorphism, but not a monomorphism. What happens if K is infinite?

3.9 Suppose that R is an infinite ring such that R/I is finite for each non-trivial ideal I. Show that R is an integral domain.

3.4   Irreducibles, primes and unique factorization domains

Throughout this section we shall suppose that R is an integral domain.

We say that an element q of R is *invertible*, or a *unit*, if it has a multiplicative inverse: that is, there is an element q^{-1} of R such that $qq^{-1}=1$. If q is a unit then since R is an integral domain its inverse q^{-1} is unique. If q_1 and q_2 are units, so is q_1q_2 (it has $q_2^{-1}q_1^{-1}$ as inverse), and so the units form a group under multiplication.

Suppose now that a is a non-zero element of R which is not a unit. We say that a *factorizes* if we can write $a=bc$, where neither b nor c is a unit. If a does not factorize we say that a is *irreducible*. Thus a is irreducible if it is not zero, not a unit, and whenever $a=bc$ then either b or c is a unit.

In this section we consider the two following questions. When can every non-zero element of R which is not a unit be expressed as a product of irreducible elements? If such factorization is possible, when is it essentially unique?

We begin by characterizing irreducibility in terms of ideals. Let PP denote the collection of proper principal ideals. We order PP by inclusion. Recall that a principal ideal (a) in R consists of all multiples of a by elements of R. If $b \in (a)$ we say that a divides b, and write $a|b$. $a|b$ if and only if $(b) \subseteq (a)$.

Theorem 3.2   *A non-zero element a of R is irreducible if and only if (a) is a maximal element of PP.*

Proof. If a is irreducible, a does not divide 1, so that $1 \notin (a)$ and (a) is proper. Suppose that $(b) \in PP$ and that $(b) \supseteq (a)$. Then $(b) \neq R$, so that b is not a unit. Also $b \mid a$, so that we can write $a = bc$. As a is irreducible and b is not a unit, c is a unit. Thus $(b) = (a)$, and (a) is maximal in PP.

Conversely suppose that (a) is maximal in PP. As (a) is proper, a is not a unit. Suppose that $a = bc$, where b is not a unit. Then $(b) \in PP$ and $(b) \supseteq (a)$, so that $(b) = (a)$, by the maximality of (a). Thus $b = af$ for some f in R, $a = afc$, and so $fc = 1$, since R is an integral domain. Thus c is a unit and a is irreducible.

We now introduce a condition which is rather technical, but which allows us to deal with the first question that we raised. We say that R satisfies the *ascending chain condition for principal ideals* (ACCPI) if whenever $I_1 \subseteq I_2 \subseteq I_3 \subseteq \cdots$ is an increasing sequence of principal ideals then there exists n such that $I_m = I_n$ for all $m \geqslant n$.

Theorem 3.3 *If R satisfies the ACCPI, every element of PP is contained in a maximal element of PP.*
Proof. If (a) is in PP, either (a) is maximal in PP or it is contained in a strictly larger ideal (a_1) in PP. This process an be repeated at most finitely often, because of the ACCPI.

Theorem 3.4 *If R satisfies the ACCPI, every non-zero element of R which is not a unit can be expressed as the product of finitely many irreducible elements of R.*
Proof. Suppose that r is a non-zero element of R which is not a unit. If r is irreducible, there is nothing to prove. Otherwise, (r) is contained in a maximal element (a_1) of PP (Theorem 3.3). Then a_1 is irreducible (Theorem 3.2) and $a_1 \mid r$. We can therefore write $r = a_1 r_1$. If r_1 is irreducible, we are finished: if not, we can repeat the argument for r_1. Continuing in this way, *either* the process terminates after a finite number of steps, in which case we are finished, *or* the process continues indefinitely: that is, there is a sequence $a_1, a_2, a_3, \ldots$ of irreducible elements of R and a sequence $r_1, r_2, r_3, \ldots$ of elements of R such that, for each n,

$$r = a_1 a_2 \ldots a_n r_n.$$

But then $r_n = a_{n+1} r_{n+1}$, so that $(r_n) \subseteq (r_{n+1})$, for each n. As a_{n+1} is irreducible, and therefore is not a unit, $(r_n) \neq (r_{n+1})$. Thus (r_n) is a strictly increasing sequence of ideals, contradicting the ACCPI. Thus the second possibility cannot arise, and the proof is complete.

We now turn to the second problem, concerning the uniqueness of factorization. Here a little care is needed. Suppose that an element r of a ring factorizes as a product $r = r_1 r_2 \ldots r_n$ of irreducible elements. Suppose that $\varepsilon_1, \varepsilon_2, \ldots, \varepsilon_n$ are units, with $\varepsilon_1 \varepsilon_2 \ldots \varepsilon_n = 1$, and that π is a permutation of $\{1, \ldots, n\}$. Then we can write $r = r'_1 \ldots r'_n$, where $r'_i = \varepsilon_i r_{\pi(i)}$. But this is *essentially* the same factorization of r.

In order to allow for this, we make the following definitions. We say that r and s are *associates* if there exists a unit ε such that $r = \varepsilon s$. We say that an integral domain R is a *unique factorization domain* if

(a) any non-zero element r of R which is not a unit can be written as a finite product of irreducible elements, and

(b) if $r_1 \ldots r_m$ and $s_1 \ldots s_n$ are two factorizations of r as products of irreducible elements then $m = n$ and there is a permutation π of $\{1, \ldots, m\}$ such that r_i and $s_{\pi(i)}$ are associates for $1 \leqslant i \leqslant m$.

If r is a non-zero element in a unique factorization domain then the number of irreducible factors (in a factorization of r as a product of irreducible elements) is called the *length* of r, and is denoted by $l(r)$. (If r is a unit, we set $l(r) = 0$.) If $r = st$, then $l(r) = l(s) + l(t)$.

In order to characterize unique factorization domains, we need to introduce a new concept. An integer n is irreducible in the integral domain $\mathbb{Z}$ if and only if $|n|$ is a prime number. In algebra it is customary to use the term 'prime' in a special way. A non-zero element a of an integral domain is said to be a *prime* if a is not a unit and whenever $a | bc$ then either $a | b$ or $a | c$.

Theorem 3.5 *A prime element of an integral domain R is irreducible.*
Proof. If a is a prime and $a = bc$ then either $a | b$ or $a | c$. If $a | b$ we can write $b = af$ for some f in R, so that $a = afc$. As R is an integral domain, $1 = fc$, and c is a unit. Similarly if $a | c$ then b is a unit.

Theorem 3.6 *An integral domain R is a unique factorization domain if and only if R satisfies the ACCPI and every irreducible element of R is a prime.*
Proof. Suppose first that R is a unique factorization domain. Suppose that $(a_1) \subseteq (a_2) \subseteq \cdots$ is an increasing sequence of principal ideals. If all the a_i are 0, the series terminates. Otherwise there is a least j such that $a_j \neq 0$. The sequence $(l(a_j), l(a_{j+1}), \ldots)$ is a decreasing sequence of non-negative integers, and so there exists n such that $l(a_m) = l(a_n)$ for all $m \geqslant n$. This means that a_m and a_n are associates, and so $(a_m) = (a_n)$ for $m \geqslant n$. Thus R satisfies the ACCPI.

Suppose that r is irreducible and that r divides ab. We can write $ab = rc$. If a is a unit, $r | b$; if b is a unit $r | a$. Otherwise we can write

$$a = s_1 \ldots s_l, \quad b = t_1 \ldots t_m, \quad c = u_1 \ldots u_n$$

as products of irreducibles. Then

$$ab = s_1 \ldots s_l t_1 \ldots t_m = r u_1 \ldots u_n,$$

and r is an associate of an s_i or a t_j. Thus r divides a or b, and r is a prime.

Conversely, suppose that the conditions are satisfied. By Theorem 3.4, any non-zero element r of R which is not a unit can be expressed as a product of irreducible elements. Let $m(r)$ be the least number of irreducible factors in any such product. We prove that condition (b) of the definition holds by induction on $m(r)$. The result certainly holds when $m(r) = 1$, for then r is irreducible. Suppose that $m > 1$, that the result holds for all s with $m(s) < m$ and that $r = r_1 \ldots r_m$ is an element with $m(r) = m$. Suppose that $r = s_1 \ldots s_n$ is another factorization of r into irreducible factors. By hypothesis, r_m is a prime and $r_m | s_1 \ldots s_n$; by repeated use of the definition of a prime element, r_m must divide s_j for some $1 \leqslant j \leqslant n$. By relabelling, we can suppose that $j = n$. Then $s_n = u r_m$ for some u in R. As s_n is irreducible, u must be a unit, and so r_m and s_n are associates. Now if $r' = (u^{-1} r_1) r_2 \ldots r_{m-1}$ then $m(r') < m$ and $r' = s_1 \ldots s_{n-1}$. By the inductive hypothesis, $m - 1 = n - 1$ and there is a permutation π of $\{1, \ldots, m-1\}$ such that $u^{-1} r_1$ and $s_{\pi(1)}$ are associates, and so are r_j and $s_{\pi(j)}$ for $2 \leqslant j \leqslant m - 1$: this establishes the induction and completes the proof.

Exercises

3.10 What are the units in $R[x]$, where R is an integral domain? What are the units in $\mathbb{Z}_4[x]$?

3.11 Show that an element a of an integral domain R is prime if and only if $R/(a)$ is an integral domain.

3.12 Let $R = \mathbb{Z} + i\sqrt{5}\,\mathbb{Z}$.

(a) Show that the units are 1 and -1.

(b) Let $\phi(m + i\sqrt{5}\,n) = m^2 + 5n^2$. $\phi(0) = 0$, $\phi(1) = \phi(-1) = 1$ and otherwise $\phi(a) > 3$. Use this to show that $2 + i\sqrt{5}$ and $2 - i\sqrt{5}$ are irreducible in R.

(c) Show that $2 + i\sqrt{5}$ is not a prime; R is not a unique factorization domain.

3.13 Show that $\mathbb{Z} + i\sqrt{5}\,\mathbb{Z}$ satisfies the ACCPI.

3.14 A proper ideal I of a ring R is said to be *prime* if whenever $ab \in I$ then either $a \in I$ or $b \in I$. Show that a non-zero element c of R is prime if and only if (c) is a prime ideal.

3.5   Principal ideal domains

Recall that a principal ideal (a) in a ring R consists of all multiples of a by elements of R.

It turns out that many important integral domains have the property that every ideal is principal. Such integral domains are called *principal ideal domains*. Let us give some examples. The ring $\mathbb{Z}$ of integers is a principal ideal domain: the sets $n\mathbb{Z}$ are ideals in $\mathbb{Z}$, and every ideal is of this form. If K is a field, the ring $K[x]$ of polynomials in one variable is a principal ideal domain. This is a consequence of the following theorem:

Theorem 3.7 *Suppose that f and g are non-zero elements of $K[x]$ (where K is a field). Then there exist elements q and r in $K[x]$ such that*

$$g = qf + r$$

and either $r = 0$ or degree $r <$ degree f.

Proof. The proof, which is by induction on degree g, is a matter of long division. If degree $f = 0$, we can take $q = f^{-1}g$ and $r = 0$, and so we need only consider the case where degree $f = k > 0$. If degree $g < k$, we can take $q = 0$ and $r = g$. Suppose that the result holds for all polynomials g of degree less than n (where $n \geqslant k$) and that

$$g = g_0 + \cdots + g_n x^n$$

has degree n. Suppose that

$$f = f_0 + \cdots + f_k x^k.$$

Let $\lambda = f_k^{-1}g_n$. Then

$$\lambda x^{n-k}f = \lambda f_0 x^{n-k} + \cdots + g_n x^n$$

so that $h = g - \lambda x^{n-k}f$ has degree less than n. By the inductive hypothesis, there exist q and r such that $h = qf + r$ and either $r = 0$ or degree $r <$ degree f. Then

$$g = (\lambda x^{n-k} + q)f + r,$$

and so the induction is established.

Theorem 3.8 *If K is a field, $K[x]$ is a principal ideal domain.*

Proof. Suppose that J is an ideal in $K[x]$ other than $\{0\}$, and let f be a non-zero polynomial of minimal degree in J. If $g \in J$, by Theorem 3.7 there exist q and r in $K[x]$ such that

$$g = qf + r$$

and either $r = 0$ or degree $r <$ degree f. But $r = g - qf \in J$, and so, by the minimality of degree f, $r = 0$. Thus $g \in (f)$. Consequently $J \subseteq (f)$. As $(f) \subseteq J$, the result is proved.

On the other hand, $K[x, y]$ is not a principal ideal domain: the ideal (x, y) is clearly not principal. Similarly $\mathbb{Z}[x]$ is not a principal ideal domain: the ideal $(2, x)$ is not principal.

Theorem 3.9 *A principal ideal domain R is a unique factorization domain.*
Proof. We use Theorem 3.6. Suppose first that $I_1 \subseteq I_2 \subseteq \cdots$ is an increasing sequence of principal ideals. Let $J = \bigcup_{j=1}^{\infty} I_j$. Then J is an ideal in R and so, since R is a principal ideal domain, $J = (a)$ for some a in J. But then $a \in I_j$ for some j, so that $J = (a) \subseteq I_j$. This clearly means that $I_k = I_j$ for all $k > j$.

Secondly suppose that a is irreducible in R and that $a | bc$. Suppose that a does not divide b. Then $b \notin (a)$, so that $(a, b) \neq (a)$. But (a) is maximal in PP (Theorem 3.2), and (a, b) is a principal ideal: it follows that $(a, b) = R$. Thus there exist p and q in R such that

$$ap + bq = 1.$$

Multiplying by c,

$$apc + bcq = c;$$

as $a | bc$, it follows that $a | c$. Thus a is prime.

Exercises

3.15 Suppose that $f = k_0 + k_1 x + \cdots + k_n x^n$ is a non-zero element of $K[x]$ (where K is a field). An element α of K is a *root* of f if $f(\alpha) = k_0 + k_1 \alpha + \cdots + k_n \alpha^n = 0$. Show that α is a root of f if and only if $f \in (x - \alpha)$, and show that f has at most n distinct roots.

3.16 What are the ideals in $\mathbb{Z}_6$? Is it a principal ideal domain?

3.17 Suppose that p is a prime number. Let R be the set of rationals which can be written in the form r/s, where p does not divide s. Show that R is a subring of $\mathbb{Q}$. What are the units in R? Show that R is a principal ideal domain.

3.18 Suppose that R is a principal ideal domain which is not a field. Show that $R[x]$ is not a principal ideal domain.

3.6 Highest common factors

Suppose that B is a subset of a ring R. We say that an element a of R is a *highest common factor* of B if first $a | b$ for each b in B and secondly if $a' | b$ for each b in B then $a' | a$. We can express this in terms of ideals: a is a highest common factor of B if first $B \subseteq (a)$ and secondly if $B \subseteq (a')$ then $(a) \subseteq (a')$.

This means that if a and a' are highest common factors of B then $(a)=(a')$: thus if R is an integral domain a and a' are associates.

Note also that, if B is a non-empty set of non-zero elements of a principal ideal domain R, the ideal (B) is principal, and so $(B)=(a)$ for some a; clearly a is a highest common factor of B. Further there exist $b_1,\ldots,b_n$ in B and $r_1,\ldots,r_n$ in R such that

$$a=r_1b_1+\cdots+r_nb_n.$$

We shall now show that the first of these properties (but not always the second: see Exercise 3.20) holds in unique factorization domains.

Theorem 3.10 *If B is a non-empty set of non-zero elements of a unique factorization domain R, B has a highest common factor.*
Proof. The idea behind the proof is simple: we consider the irreducible common factors of B, and combine as many as possible to obtain a highest common factor. Let

$$D=\{r\in R: r\,|\,b \text{ for each } b \text{ in } B\}$$

be the set of common factors of B. D is non-empty, since $1\in D$. If $r\in D$ and $b\in B$, $l(r)\leqslant l(b)$, and so D contains an element a of maximal length. We shall show that a is a highest common factor of B.

Suppose that a' is another element of D. Among the common factors of a and a' there is one of maximal length, c say. We can write

$$a=cd, \quad a'=cd'.$$

Suppose that d' is not a unit. Let π' be an irreducible factor of d'. Then $l(c\pi')=l(c)+1$, so that $c\pi'$ does not divide a, since $l(c)$ is as large as possible. This means that π' does not divide d. Now if b is any element of B, we can write $b=af=cdf$. As $a'\,|\,b$, $c\pi'\,|\,b$ and so $\pi'\,|\,df$. π' does not divide d, and π' is a prime (by Theorem 3.6), so that $\pi'\,|\,f$. Thus $a\pi'\,|\,b$. This holds for every b in B, and so $a\pi'\in D$. But $l(a\pi')=l(a)+1$, contradicting the maximality of $l(a)$. As a consequence, d' must be a unit, and so $a'\,|\,a$.

Suppose that B is a non-empty subset of a ring R. We say that B is *relatively prime* if 1 is a highest common factor of B. If a is a highest common factor of a non-empty subset B of an integral domain R then the set $\{c: ca\in B\}$ is relatively prime.

Exercises

3.19 Suppose that R is an integral domain. Show that the following are equivalent:
(i) every finite non-empty set of non-zero elements of R has a highest common factor;

(ii) every finite non-empty set of non-zero elements of R has a least common multiple.

3.20 Show that 1 and -1 are the highest common factors of 2 and x in $\mathbb{Z}[x]$ and that neither can be written in the form $2a + xb$, with a and b in $\mathbb{Z}[x]$.

3.21 Suppose that R is an integral domain with the property that every non-empty set B of non-zero elements has a highest common factor of the form $\gamma_1 b_1 + \cdots + \gamma_n b_n$, with $b_1, \ldots, b_n$ in B and $\gamma_1, \ldots, \gamma_n$ in R. Show that R is a principal ideal domain.

3.7 Polynomials over unique factorization domains

Galois theory is largely concerned with polynomials in one variable, with coefficients in a field K. We shall, however, also need to consider polynomials with integer coefficients, and to consider polynomials in several variables. In order to deal with both of these, it is convenient to study polynomial rings of the form $R[x]$, where R is a unique factorization domain.

In this section, we shall suppose that R is a unique factorization domain, with field of fractions F. If

$$f = a_0 + a_1 x + \cdots + a_n x^n$$

is a non-zero element of $R[x]$, we define the *content* of f to be a highest common factor of the non-zero coefficients of f (the fact that this is not uniquely defined causes no problems). If f has content 1 we say that f is *primitive*. If γ is the content of f then $f = \gamma g$, where g is primitive.

If f is an element of $R[x]$, we can consider f as an element of $F[x]$. The next theorem provides a partial converse.

Theorem 3.11 *Suppose that R is a unique factorization domain. An element of $R[x]$ is a unit if and only if it is a unit in R. If f is a non-zero element of $F[x]$ we can write $f = \beta g$, where g is a primitive polynomial in $R[x]$ and $\beta \in F$. If $f = \beta' g'$ is another such expression then g and g' are associates in $R[x]$; there exists a unit ε in R such that $g = \varepsilon g'$.*

Proof. The first statement is obvious.

Suppose that f is a non-zero element of $F[x]$. We clear denominators: there exists δ in R such that $\delta f \in R[x]$. Let γ be the content of δf. Then $\delta f = \gamma g$, where g is primitive in $R[x]$, and so $f = (\delta^{-1}\gamma)g = \beta g$.

Suppose that $f = \beta' g'$ is another such expression. We again clear denominators: there exists α in R such that $\alpha\beta$ and $\alpha\beta'$ are in R. Then $\alpha f = (\alpha\beta)g = (\alpha\beta')g'$. As g is primitive in $R[x]$, $\alpha\beta$ is the content of αf: so is $\alpha\beta'$ (remember that the content is not uniquely defined!), and so $\alpha\beta$ and $\alpha\beta'$ are associates in R. This means that there is a unit in R such that $\alpha\beta' = \varepsilon\alpha\beta$: $\alpha\beta g = \alpha\beta' g' = \varepsilon\alpha\beta g'$, so that $g = \varepsilon g'$ and g and g' are associates in $R[x]$.

Theorem 3.12 *Suppose that R is a unique factorization domain. If f and g are primitive elements of $R[x]$, so is fg.*
Proof. Suppose that

$$f = a_0 + a_1 x + \cdots + a_n x^n,$$
$$g = b_0 + b_1 x + \cdots + b_m x^m,$$

and

$$fg = c_0 + c_1 x + \cdots + c_{m+n} x^{m+n}.$$

Let d be the content of fg and suppose that d is not a unit. Let r be an irreducible factor of d. As R is a unique factorization domain, r is a prime. Since f is primitive, there exists a least i such that r does not divide a_i; similarly there exists a least j such that r does not divide b_j. As r is a prime, r does not divide $a_i b_j$. We consider the coefficient

$$c_{i+j} = \sum_{k<i} a_k b_{i+j-k} + a_i b_j + \sum_{l<j} a_{i+j-l} b_l.$$

Now r divides a_k for $k < i$, and so r divides $\sum_{k<i} a_k b_{i+j-k}$; similarly r divides $\sum_{l<j} a_{i+j-l} b_l$. But r also divides c_{i+j}, and so r divides $a_i b_j$: this gives the required contradiction.

Corollary (Gauss' lemma) *An element g of $R[x]$ is irreducible if and only if either it is an irreducible element of R or it is primitive, and irreducible in $F[x]$.*
Proof. Suppose that g is irreducible in $R[x]$. If degree $g = 0$, then g must be irreducible in R. If degree $g > 0$, then g must certainly be primitive. Suppose that $g = f_1 f_2$ is a factorization in $F[x]$. By Theorem 3.11, we can write $f_1 = \beta_1 g_1$, $f_2 = \beta_2 g_2$ with β_1 and β_2 in F, and g_1 and g_2 primitive in $R[x]$. Thus

$$g = \beta_1 \beta_2 g_1 g_2$$

Now $g_1 g_2$ is primitive, so that, by Theorem 3.11, $\beta_1 \beta_2$ is a unit in R. Thus $g = (\beta_1 \beta_2 g_1) g_2$, contradicting the irreducibility of g.

The converse implications are clear.

We now come to the main result of this section.

Theorem 3.13 *If R is a unique factorization domain, so is $R[x]$.*
Proof. Suppose that f is a non-zero element of $R[x]$. Then $f = \alpha g$, where α is the content of f and g is primitive in $R[x]$. We now consider g as an element of $F[x]$. $F[x]$ is a unique factorization domain, and so we can write $g =$

$g_1 \ldots g_k$ as a product of irreducible elements in $F[x]$. By Theorem 3.11, we can write each g_j as $\beta_j f_j$, where $\beta_j \in F$ and f_j is primitive in $R[x]$. Note that each f_j is irreducible in $R[x]$, by Gauss' lemma. Thus

$$g = \beta f_1 \ldots f_k,$$

where $\beta = \beta_1 \ldots \beta_k$. By Theorem 3.12, $f_1 \ldots f_k$ is primitive, and so β is a unit in R, by Theorem 3.11. Thus we can write

$$f = \alpha_1 \ldots \alpha_j f_1 \ldots f_k$$

where $\alpha_1 \ldots \alpha_j$ is a factorization of $\alpha\beta$ as a product of irreducible elements of R. Thus f can be expressed as a product of irreducible elements of $R[x]$.

Suppose that

$$f = \alpha_1' \ldots \alpha_l' f_1' \ldots f_m'$$

is another such factorization. As $\alpha_1 \ldots \alpha_j$ and $\alpha_1' \ldots \alpha_l'$ are both contents of f, they are associates in R; since R is a unique factorization domain, $l = j$ and there exists a permutation π of $\{1, \ldots, j\}$ such that α_i and $\alpha_{\pi(i)}'$ are associates for $1 \leqslant i \leqslant j$.

Further, $f_1 \ldots f_k = \lambda f_1' \ldots f_m'$, where λ is a unit in R. By Gauss' lemma, $f_1, \ldots, f_k, f_1', \ldots, f_m'$ are irreducible in $F[x]$. As $F[x]$ is a unique factorization domain, $m = k$ and there exists a permutation ρ of $\{1, \ldots, k\}$ and non-zero elements $\varepsilon_1, \ldots, \varepsilon_k$ of F such that $f_i = \varepsilon_i f_{\rho(i)}'$ for $1 \leqslant i \leqslant k$. But f_i and $f_{\rho(i)}'$ are primitive in $R[x]$, by Gauss' lemma, and so f_i and $f_{\rho(i)}'$ are associates in $R[x]$, by Theorem 3.11. Thus $R[x]$ is a unique factorization domain.

Corollary 1 *Suppose that f is a primitive element of $R[x]$, that g is a non-zero element of $R[x]$ and that f divides g in $F[x]$. Then f divides g in $R[x]$.*
Proof. We can factorize g as

$$g = \alpha_1 \ldots \alpha_j g_1 \ldots g_k$$

where the α_i are irreducible in R and the g_i are irreducible elements of $R[x]$ of positive degree. By Gauss' lemma, each g_i is primitive and irreducible in $F[x]$. Thus

$$g = (\alpha_1 \ldots \alpha_j g_1) g_2 \ldots g_k$$

is a factorization of g as a product of irreducible elements of $F[x]$. As f divides g in $F[x]$, and as $F[x]$ is a unique factorization domain, we can write

$$f = \varepsilon g_{i_1} \ldots g_{i_r}$$

where ε is a non-zero element of F and $1 \leqslant i_1 < \ldots < i_r \leqslant k$. Now $g_{i_1} \ldots g_{i_r}$ is primitive, by Theorem 3.12. As f is also primitive, ε is a unit in R, by Theorem 3.11, and so f divides g in $R[x]$.

Corollary 2 *If R is a unique factorization domain, then so is $R[x_1, \ldots, x_n]$.*

Exercises

3.22 Express all the cubic polynomials (polynomials of degree 3) in $\mathbb{Z}_2[x]$ as products of irreducible factors.

3.23 Express all the homogeneous quadratic polynomials (polynomials of degree 2 with no constant or linear terms) in $\mathbb{Z}_2[x, y, z]$ as products of irreducible factors.

3.8   The existence of maximal proper ideals

By Theorem 3.2, a non-zero element a of a principal ideal domain R which is not a unit is irreducible if and only if (a) is a maximal proper ideal in R. This suggests that maximal ideals are important: Theorem 3.16 in the next section shows that this is indeed so. Are there many such ideals?

Theorem 3.14 *Suppose that J is a proper ideal of a ring R. Then there exists a maximal proper ideal which contains J.*
Proof. We use Zorn's lemma. Let P_J denote the collection of proper ideals of R which contain J. We order P_J by inclusion. If C is a chain in P_J, let

$$I_0 = \bigcup \{I : I \in C\}.$$

If a_1 and $a_2 \in I_0$, there exist I_1 and I_2 in C such that $a_1 \in I_1$ and $a_2 \in I_2$. As C is a chain, either $I_1 \subseteq I_2$ or $I_2 \subseteq I_1$. If $I_1 \subseteq I_2$, then $a_1 \in I_2$ and so $a_1 + a_2 \in I_2$. As $I_2 \subseteq I_0$, $a_1 + a_2 \in I_0$. A similar argument applies if $I_2 \subseteq I_1$. More trivially, if $a \in I_0$ and $r \in R$, then $a \in I_1$ for some I_1 in C. Then $ra \in I_1$, and so $ra \in I_0$. Thus I_0 is an ideal.

Further, if $I \in C$, $1 \notin I$, since I is proper. Thus $1 \notin I_0$. Consequently $I_0 \in P_J$; I_0 is an upper bound for C, and so P_J contains a maximal element, by Zorn's lemma. Finally, it is obvious that a maximal element of P_J is also a maximal proper ideal in R.

The generality of Theorem 3.14 suggests that the use of the axiom of choice is natural here. This is indeed so: if one assumes the truth of Theorem 3.14, one can deduce the axiom of choice. The problem that follows shows that the full force of the axiom of choice is frequently not needed.

Exercises

3.24 Prove Theorem 3.14 in the case where R has countably many elements, without using Zorn's lemma.

3.25 Show that an element of a ring R is invertible if and only if it is contained in no maximal proper ideal in R.

3.26 Suppose that J is a proper prime ideal in an integral domain R. Show that $J[x]$ is prime in $R[x]$. Show that $J[x]$ is not a maximal proper ideal in $R[x]$.

3.9 More about fields

A field is a ring in which every non-zero element has an inverse. How can we recognize when a ring is a field?

Theorem 3.15 *A ring R is a field if and only if $\{0\}$ and R are the only ideals in R.*

Proof. Suppose first that R is a field, that I is an ideal in R other than $\{0\}$ and that a is a non-zero element of I. If b is any element of R, $b = a(a^{-1}b) \in I$, and so $I = R$.

Conversely, suppose that R is a ring whose only ideals are $\{0\}$ and R. If a is a non-zero element of R, the principal ideal (a) must be R, and so there exists b in R such that $ab = 1$; consequently R is a field.

Note that if ϕ is a ring homomorphism from a field K into a ring, the kernel of ϕ is a proper ideal of K, and so ϕ is one–one.

Theorem 3.15 makes it easy to decide when a quotient ring is a field.

Theorem 3.16 *If J is a proper ideal in a ring R then R/J is a field if and only if J is a maximal proper ideal in R.*

Proof. Let q denote the quotient map $R \to R/J$. If I is a proper ideal in R/J then $q^{-1}(I)$ is a proper ideal in R; further $q^{-1}(I) = J$ if and only if $I = \{0\}$. It follows from this that if J is a maximal proper ideal then $\{0\}$ and R/J are the only ideals in R/J, and R/J is a field, by Theorem 3.15.

Suppose conversely that R/J is a field. If $a \notin J$, $q(a) \neq 0$, and so, since q is onto, there exists b in R such that $q(b) = (q(a))^{-1}$. Thus

$$q(ab - 1_R) = q(a)q(b) - 1_{R/J} = 0$$

so that $ab - 1_R \in J$. There therefore exists j in J such that $1_R = ab + j$, and so $(J \cup \{a\}) = R$. This means that J is a maximal proper ideal.

Combining this with Theorem 3.2, we obtain the following.

Corollary *If a is a non-zero non-unit element of a principal ideal domain R, $R/(a)$ is a field if and only if a is irreducible.*

Applying this to the ring of integers, we see that $\mathbb{Z}_n$ is a field if and only if n is a prime number.

Suppose now that K is a field. A *subfield* of K is a subset of K which is a field under the operations inherited from K. Any subfield contains 0 and 1. The intersection of all subfields is again a subfield, the smallest subfield of K. This subfield is called the *prime subfield* of K.

K is a ring; we can consider the homomorphism ϕ from $\mathbb{Z}$ into K described in Section 3.3. If K has non-zero characteristic n then, since K is certainly an integral domain, n must be a prime number. Thus $\phi(\mathbb{Z})$, which is isomorphic to $\mathbb{Z}_n$, is a field. Clearly it is the prime subfield of K.

The other possibility is that K has characteristic 0. In this case $\phi(\mathbb{Z})$ is a subring of K isomorphic to $\mathbb{Z}$. If $q = r/s$ is a rational, let us define $\phi(q)$ by setting $\phi(q) = \phi(r)\phi(s)^{-1}$. If r'/s' is another expression for q, $rs' = r's$, so that

$$\phi(r)\phi(s') = \phi(r')\phi(s)$$

and so

$$\phi(r)\phi(s)^{-1} = \phi(r')\phi(s')^{-1}.$$

Thus ϕ is properly defined, and it is equally straightforward to verify that ϕ is a ring homomorphism of $\mathbb{Q}$ into K. $\phi(\mathbb{Q})$ is a subfield of K. Clearly every element of $\phi(\mathbb{Q})$ is in every subfield of K, so that $\phi(\mathbb{Q})$ is the prime subfield of K. Summing up:

Theorem 3.17 *Suppose that K is a field. If K has characteristic 0, the prime subfield of K is isomorphic to $\mathbb{Q}$. Otherwise, K has prime characteristic, p say, and the prime subfield of K is isomorphic to $\mathbb{Z}_p$.*

Exercise

3.27 Let R be the ring of Exercise 3.17. What are the possible quotient fields of R?

PART 2

The theory of fields, and Galois theory

4

Field extensions

4.1 Introduction

One of the main topics of Galois theory is the study of polynomial equations. In order to consider how we should proceed, let us first consider some rather trivial and familiar examples.

Polynomials involve addition and multiplication, and so it is natural to consider polynomials with coefficients in a ring R. If we consider the simplest possible case, when $R = \mathbb{Z}$ and p is a polynomial of degree 1, we find there are difficulties: for example, we cannot solve the equation $2x + 3 = 0$ in $\mathbb{Z}$.

In the case where R is an integral domain, the field of fractions is constructed in order to deal with this problem. Thus, in the example above, if we consider 2 and 3 as elements of $\mathbb{Q}$, the rational field, the equation has a solution $x = -3/2$ in $\mathbb{Q}$.

Let us now consider a quadratic equation: $x^2 - 2x - 1 = 0$. We consider this as an equation with rational coefficients: completing the square, we find that

$$(x-1)^2 = 2.$$

As we have seen in Section 1.2, there is no rational number r such that $r^2 = 2$, so the quadratic equation has no solution in $\mathbb{Q}$. Instead, the first natural idea is to consider the polynomial as a polynomial with *real* coefficients: the equation then factorizes as $(x - 1 + \sqrt{2})(x - 1 - \sqrt{2}) = 0$, and we have solutions $1 - \sqrt{2}$ and $1 + \sqrt{2}$.

The field $\mathbb{R}$ is rather large, however ($\mathbb{R}$ is uncountable, while $\mathbb{Q}$ is countable), and it is possible to proceed more economically. Recall that in Section 1.2 we showed that the set of all numbers of the form $r + s\sqrt{2}$, where r and s are rationals, forms a field K. Clearly $\mathbb{R} \supseteq K \supseteq \mathbb{Q}$, and K appears to be much smaller than R. If we consider $x^2 - 2x - 1$ as an element of $K[x]$, we can solve the equation in K.

Let us express all this in more algebraic language. The polynomial $x^2 - 2x - 1$ is irreducible in $\mathbb{Q}[x]$ (and is therefore irreducible in $\mathbb{Z}[x]$). If, however, it is considered as an element of $K[x]$ or $\mathbb{R}[x]$, it can be written as a product of linear factors. This suggests the following general programme: given an element f in $K[x]$ (where K is a field), can we find a larger field, L say, such that f considered as an element of $L[x]$ can be written as a product of linear factors? If so, can we do it in an economical way?

4.2   Field extensions

Suppose that we start with a field K. In order to construct a larger field L we frequently have, by some means or another, to construct L, and then find a subfield of L which is isomorphic to K (think of how the complex numbers are constructed from the reals). It is occasionally important to realize that this sort of procedure is adopted: for this reason we define an *extension* of a field K to be a triple (i, K, L), where L is another field, and i is a (ring) monomorphism of K into L.

On the other hand, much more frequently this is far too cumbersome. If (i, K, L) is an extension of K, the image $i(K)$ is a subfield of L which is isomorphic to K; we shall usually identify K with $i(K)$ and consider it as a subfield of L. In this case we shall write $L:K$ for the extension. Thus $\mathbb{C}:\mathbb{R}$ is the extension of the real numbers by the complex numbers and $\mathbb{R}:\mathbb{Q}$ is the extension of the rational numbers by the real numbers. Very occasionally, when the going gets rough, we shall need to be rather careful: in these circumstances we shall revert to the notation (i, K, L).

Suppose now that $L:K$ is an extension. How do we measure how big the extension is? It turns out that the appropriate idea is dimension, in the vector space sense. If you are reasonably familiar with the idea of the dimension of a vector space (as you should be) you will find this an almost embarrassingly simple idea: the remarkable thing is that it is extraordinarily powerful.

To begin with, then, we forget about many of the field properties of L.

Theorem 4.1 *Suppose that $L:K$ is an extension. Under the operations*

$$(l_1, l_2) \to l_1 + l_2 \ \textit{from } L \times L \textit{ to } L$$

and

$$(k, l) \to kl \ \textit{from } K \times L \textit{ to } L,$$

L is a vector space over K.
Proof. All the axioms are satisfied.

Thus $\mathbb{C}$ is a real vector space, and $\mathbb{R}$ is a vector space over the rationals $\mathbb{Q}$.

We now define the *degree of an extension* $L:K$ to be the dimension of L as a vector space over K. We write $[L:K]$ for the degree of $L:K$. We say that $L:K$ is *finite* if $[L:K] < \infty$, and that $L:K$ is *infinite* if $[L:K] = \infty$.

Thus $[\mathbb{C}:\mathbb{R}] = 2$, $[\mathbb{R}:\mathbb{Q}] = \infty$, and, if K is the field of all $r + s\sqrt{2}$, with r and s rational, $[K:\mathbb{Q}] = 2$. In this sense, then, $K:\mathbb{Q}$ is a more economical extension for solving $x^2 - 2x - 1 = 0$ than $\mathbb{R}:\mathbb{Q}$.

The next theorem is very straightforward (there is an obvious argument to try, and it works), but it is the key to much that follows. If $M:L$ and $L:K$ are extensions, then clearly so is $M:K$.

Theorem 4.2 *Suppose that $M:L$ and $L:K$ are extensions. Then*
$$[M:K] = [M:L][L:K].$$
Proof. First suppose that the right-hand side is finite, so that we can write $[M:L] = m < \infty$, and $[L:K] = n < \infty$. Let $(x_1, \ldots, x_m)$ be a basis for M over L, and let $(y_1, \ldots, y_n)$ be a basis for L over K. We can form the products $y_j x_i$ (for $1 \leqslant i \leqslant m$, $1 \leqslant j \leqslant n$) in M. We shall show that the mn elements $(y_j x_i : 1 \leqslant i \leqslant m, 1 \leqslant j \leqslant n)$ form a basis for M over K.

First we show that they span M over K. Let $z \in M$. As $(x_1, \ldots, x_m)$ is a basis for M over L, there exist $\alpha_1, \ldots, \alpha_m$ in L such that
$$z = \alpha_1 x_1 + \cdots + \alpha_m x_m.$$
As each α_i is in L, and as $(y_1, \ldots, y_n)$ is a basis for L over K, for each i there exist $\beta_{i1}, \ldots, \beta_{in}$ in K such that
$$\alpha_i = \beta_{i1} y_1 + \cdots + \beta_{in} y_n.$$
Substituting,
$$z = \sum_{i=1}^{m} \sum_{j=1}^{n} \beta_{ij} y_j x_i$$
which proves our assertion.

Secondly we show that $(y_j x_i : 1 \leqslant i \leqslant m; 1 \leqslant j \leqslant n)$ is a linearly independent set over K. Suppose that
$$0 = \sum_{i=1}^{m} \sum_{j=1}^{n} \gamma_{ij} y_j x_i$$
where the γ_{ij} are elements of K. Let us set
$$\delta_i = \sum_{j=1}^{n} \gamma_{ij} y_j \; (\in L)$$
for $1 \leqslant i \leqslant m$. Then
$$0 = \sum_{i=1}^{m} \delta_i x_i.$$

But $(x_1, \ldots, x_m)$ is linearly independent over L, and so $\delta_i = 0$ for $1 \leqslant i \leqslant m$; that is,

$$0 = \sum_{j=1}^{n} \gamma_{ij} y_j, \text{ for } 1 \leqslant i \leqslant m.$$

Now $\gamma_{ij} \in K$, and $(y_1, \ldots, y_n)$ is a linearly independent set over K. Consequently $\gamma_{ij} = 0$ for all i and j, and the second assertion is proved. Thus the elements $(y_j x_i)_{i=1, j=1}^{m, n}$ form a basis for M over K, and

$$[M:K] = [M:L][L:K]$$

provided that the right-hand side is finite.

If $[M:K] = l < \infty$, we can find a basis $(z_1, \ldots, z_l)$ for M over K. $(z_1, \ldots, z_l)$ spans M over K, and so it certainly spans M over L. Thus $[M:L] < \infty$. Also L is a K-linear subspace of M, so that $[L:K] < \infty$ (by Theorem 1.5). Thus, if the right-hand side is infinite, we must have $[M:K] = \infty$: the proof is complete.

We can extend this result in an obvious way. A sequence $K_n:K_{n-1}$, $K_{n-1}:K_{n-2}, \ldots, K_1:K_0$ of extensions, where each field extends its successor, is called a *tower*. Clearly

$$[K_n:K_0] = [K_n:K_{n-1}][K_{n-1}:K_{n-2}] \ldots [K_1:K_0];$$

we refer to this (and to Theorem 4.2) as the *tower law* for field extensions.

Exercise

4.1 Suppose that $[L:K]$ is a prime number. What fields are there intermediate between L and K?

4.3   Algebraic and transcendental elements

Suppose that $L:K$ is an extension, and that A is a subset of L. We write $K(A)$ for the intersection of all subfields of L which contain K and A. $K(A)$ is a subfield of L, and is the smallest subfield of L containing K and A. Clearly $L:K(A)$ and $K(A):K$ are extensions. $K(A):K$ is the *extension of K generated by A*.

It is useful to see what a typical element of $K(A)$ looks like. Let

$$S = \{\alpha_1 \ldots \alpha_k : \alpha_i \in A \cup \{1\}\}$$

be the set of all finite products of elements of A, together with 1, let V be the K-linear subspace of L generated by S and let $V^* = V \backslash \{0\}$. Then

$$K(A) = \{rs^{-1} : r \in V, s \in V^*\};$$

for clearly anything in the right-hand side belongs to $K(A)$, and it is a straightforward matter to verify that the right-hand side is a subfield of L containing K and A.

If $A = \{\alpha_1, \ldots, \alpha_n\}$, we write $K(\alpha_1, \ldots, \alpha_n)$ for $K(A)$. In particular, we say that an extension $L:K$ is *simple* if there exists α in L such that $L = K(\alpha)$. Thus $\mathbb{C}:\mathbb{R}$ is a simple extension, since $\mathbb{C} = \mathbb{R}(i)$. Similarly, the field K of all $m + n\sqrt{2}$, with m and n in $\mathbb{Q}$, is $\mathbb{Q}(\sqrt{2})$.

It follows from the description of $K(A)$ that, if $L:K$ is a simple extension of K and if K is countable, then L is also countable; thus $\mathbb{R}:\mathbb{Q}$ is not a simple extension.

Suppose now that $L:K$ is an extension and that $\alpha \in L$. There are two possibilities. First, there may be a non-zero polynomial $f = k_0 + k_1 x + \cdots + k_n x^n$ in $K[x]$ such that

$$f(\alpha) = k_0 + k_1 \alpha + \cdots + k_n \alpha^n = 0.$$

In other words, α is a *root* of f. In this case we say that α is *algebraic* over K. Secondly, it may happen that no such polynomial exists: in this case we shall say that α is *transcendental* over K. The two possibilities lead to very different developments: for the time being we shall concentrate on algebraic elements, and shall consider transcendental elements at a much later stage (Chapter 18).

At this point, let us remark that the study of transcendental numbers – that is, elements of $\mathbb{R}$ or $\mathbb{C}$ which are transcendental over $\mathbb{Q}$ – is one of the most difficult and profound areas of number theory. It was not until 1844 that Liouville showed that any transcendental numbers exist: this helps us to understand why Cantor's set theory, which shows that there are uncountably many transcendental numbers (see Exercise 4.7), came as such a shock. Cantor's result is of no help in particular cases: Hermite's result that e is transcendental was proved in 1873, the year before Cantor's result, and the fact that π is transcendental was proved by Lindemann in 1882. The proofs are analytical, and far away from the material of this book. For an account of transcendental number theory, see the book by Baker[1].

Let us express these ideas in terms of mappings. Suppose that $L:K$ is an extension and that $\alpha \in L$. We define the *evaluation map* E_α from $K[x]$ into L by setting $E_\alpha(f) = f(\alpha)$ for each f in $K[x]$. Notice that E_α is a ring homomorphism from $K[x]$ into L. It then follows immediately from the definitions that α is transcendental over K if and only if E_α is one–one and that α is algebraic over K if and only if E_α is not one–one.

Suppose that α is algebraic over K. The kernel K_α of the evaluation map E_α is a non-zero ideal in $K[x]$; as $K[x]$ is a principal ideal domain, there is a non-zero polynomial m_α such that $K_\alpha = (m_\alpha)$. Further, since the non-zero elements of K are the units in $K[x]$, we can take m_α to be *monic* (that is, m_α

---

[1] A. Baker, *Transcendental Number Theory*, Cambridge University Press, 1979.

has leading coefficient 1:

$$m_\alpha = k_0 + k_1 x + \cdots + k_{n-1} x^{n-1} + x^n,$$

and then m_α is uniquely determined. The polynomial m_α is called the *minimal polynomial* of α.

Theorem 4.3 *Suppose that $L:K$ is an extension and that $\alpha \in L$ is algebraic. Then m_α is irreducible in $K[x]$, the image $E_\alpha(K[x])$ of the polynomial ring $K[x]$ is the subfield $K(\alpha)$ of L, and we can factorize E_α as $i\tilde{E}_\alpha q$:*

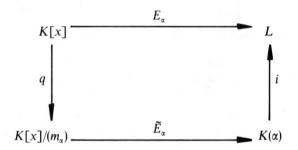

where q is the quotient mapping, $\tilde{E}_\alpha$ is an isomorphism and i is the inclusion mapping.

Proof. Suppose that $m_\alpha = fg$. Then

$$0 = E_\alpha(m_\alpha) = E_\alpha(f)E_\alpha(g) = f(\alpha)g(\alpha),$$

so that either $f(\alpha) = 0$ or $g(\alpha) = 0$. If $f(\alpha) = 0$, $f \in (m_\alpha)$, so that $m_\alpha | f$ and g is a unit. Similarly if $g \in (m_\alpha)$, f is a unit. Thus m_α is irreducible. The corollary to Theorem 3.16 implies that $K[x]/(m_\alpha)$ is a field. Now by Theorem 3.1 we can factorize E_α in the following way:

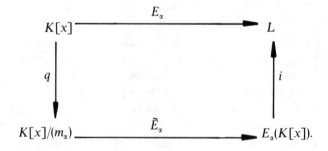

Since $\tilde{E}_\alpha$ is an isomorphism, this means that $E_\alpha(K[x])$ is a subfield of L. Since $E_\alpha(k) = k$ if $k \in K$ and $E_\alpha(x) = \alpha$, $E_\alpha(K[x]) \supseteq K \cup \{\alpha\}$, and so $E_\alpha(K[x]) \supseteq K(\alpha)$. But clearly $E_\alpha(K[x]) \subseteq K(\alpha)$, and so the proof is complete.

Let us now relate these ideas to the degree of an extension.

Theorem 4.4 *Suppose that $L:K$ is an extension and that $\alpha \in L$. Then α is algebraic over K if and only if $[K(\alpha):K] < \infty$. If this is so, then $[K(\alpha):K]$ is the degree of m_α.*

Proof. First, suppose that $[K(\alpha):K] = n < \infty$. Consider the $n+1$ terms $1, \alpha, \alpha^2, \ldots, \alpha^n$ in $K(\alpha)$. Either two terms α^r and α^s (with $0 \leqslant r < s \leqslant n$) are equal, in which case $x^s - x^r$ is in the kernel K_α of the evaluation map E_α, or they are all distinct. In this latter case, by Corollary 1 to Theorem 1.4, $\{1, \alpha, \ldots, \alpha^n\}$ are linearly dependent over K. Thus there exist $k_0, k_1, \ldots, k_n$, not all zero, such that $k_0 + k_1 \alpha + \cdots + k_n \alpha^n = 0$. Then

$$f = k_0 + k_1 x + \cdots + k_n x^n \in K_\alpha$$

so that in either case E_α is not one–one.

Next suppose that α is algebraic over K, and that m_α is the minimal polynomial of α. We shall show that if $n = \text{degree}\,(m_\alpha)$ then $\{1, \alpha, \ldots, \alpha^{n-1}\}$ forms a basis for $K(\alpha)$ over K. First we show that $\{1, \alpha, \ldots, \alpha^{n-1}\}$ is a linearly independent set over K. For if

$$k_0 \cdot 1 + k_1 \alpha + \cdots + k_{n-1} \alpha^{n-1} = 0,$$

let us set $f = k_0 + k_1 x + \cdots + k_{n-1} x^{n-1}$. Then $f \in K_\alpha = (m_\alpha)$ and degree $f <$ degree m_α, so that $f = 0$, and $k_0 = k_1 = \cdots = k_{n-1} = 0$. Secondly we show that $\{1, \alpha, \ldots, \alpha^{n-1}\}$ spans $K(\alpha)$. By Theorem 4.3, if $\beta \in K(\alpha)$ then $\beta = E_\alpha(f)$ for some $f \in K[x]$. We can write

$$f = m_\alpha q + r$$

where $r = 0$ or degree $r < n$. Then $\beta = E_\alpha(f) = E_\alpha(m_\alpha) E_\alpha(q) + E_\alpha(r) = E_\alpha(r)$ so that if $r = k_0 + k_1 x + \cdots + k_{n-1} x^{n-1}$,

$$\beta = k_0 + k_1 \alpha + \cdots + k_{n-1} \alpha^{n-1} \in \text{span}\,(1, \alpha, \ldots, \alpha^{n-1}).$$

Exercises

4.2 Suppose that $L:K$ and that K_1 and K_2 are two intermediate fields such that $L = K(K_1, K_2)$. Show that $[L:K] \leqslant [K_1:K][K_2:K]$.

4.3 Suppose that $K(\alpha):K$ is a finite simple extension. For each β in $K(\alpha)$, let $T_\alpha(\beta) = \alpha\beta$. T_α is a linear mapping of $K(\alpha)$ (considered as a vector space over K) into itself. Show that $\det(xI - T_\alpha)$ is the minimal polynomial of α over K.

4.4 Show that $x^3 + 3x + 1$ is irreducible in $\mathbb{Q}[x]$. Suppose that α is a root of $x^3 + 3x + 1$ in $\mathbb{C}$. Express α^{-1} and $(1+\alpha)^{-1}$ as linear combinations, with rational coefficients, of 1, α and α^2.

4.5 Suppose that $L(\alpha):L:K$ and that $[L(\alpha):L]$ and $[L:K]$ are relatively prime. Show that the minimal polynomial of α over L has its coefficients in K.

4.6 Suppose that $[L:K]$ is a prime number. Show that $L:K$ is simple.

4.4   Algebraic extensions

Theorem 4.4 has the following important consequence.

Theorem 4.5 *Suppose that $L:K$ is an extension. The set L_a of those elements of L which are algebraic over K is a subfield of L.*

Proof. Suppose that α and β are in L_a. As β is algebraic over K, β is certainly algebraic over $K(\alpha)$. As $K(\alpha)(\beta) = K(\alpha, \beta)$, we have $[K(\alpha, \beta):K(\alpha)] < \infty$, by Theorem 4.4. Also $[K(\alpha):K] < \infty$, by Theorem 4.4, so that, by Theorem 4.2,

$$[K(\alpha, \beta):K] = [K(\alpha, \beta):K(\alpha)][K(\alpha):K] < \infty.$$

Now $K(\alpha + \beta) \subseteq K(\alpha, \beta)$, and so $[K(\alpha + \beta):K] < \infty$. Using Theorem 4.4 again, we see that $\alpha + \beta$ is algebraic over K. Similarly $\alpha\beta$ is algebraic over K. Finally, if α is a non-zero element of L_a, with minimal polynomial

$$f = k_0 + k_1 x + \cdots + k_{n-1} x^{n-1} + x^n$$

let $g = 1 + k_{n-1} x + \cdots + k_0 x^n$. Then $g(\alpha^{-1}) = \alpha^{-n} f(\alpha) = 0$ so that α^{-1} is algebraic over K.

This theorem gives some indication of how useful the idea of the degree of an extension is. We have shown that if α and β are algebraic over K then so are $\alpha + \beta$ and $\alpha\beta$, but we have not had to produce polynomials in $K[x]$ of which these elements are roots.

We say that an extension $L:K$ is *algebraic* if every element of L is algebraic over K. Not every algebraic extension is finite: for example, if A denotes the *algebraic numbers*, the set of complex numbers which are algebraic over $\mathbb{Q}$, then $A:\mathbb{Q}$ is infinite (see Exercise 5.8 below). Finite extensions are characterized in the following way:

Theorem 4.6 *Suppose that $L:K$ is an extension. The following are equivalent:*

    (i) $[L:K] < \infty$;

    (ii) $L:K$ *is algebraic, and L is finitely generated over K;*

    (iii) *there exist finitely many algebraic elements $\alpha_1, \ldots, \alpha_n$ of L such that $L = K(\alpha_1, \ldots, \alpha_n)$.*

Proof. Suppose first that $L:K$ is finite. If $\alpha \in L$, then $[K(\alpha):K] \leqslant [L:K] < \infty$, so that α is algebraic over K (Theorem 4.4); thus $L:K$ is algebraic. If $(\beta_1, \ldots, \beta_r)$ is a basis for L over K, then $L = K(\beta_1, \ldots, \beta_r)$, so that L is finitely generated over K. Thus (i) implies (ii), and (ii) trivially implies (iii).

Suppose now that (iii) holds. Let $K_0 = K$, and let $K_j = K(\alpha_1, \ldots, \alpha_j) = K_{j-1}(\alpha_j)$, for $1 \leqslant j \leqslant n$. Note that $L = K_n$. Each α_j is algebraic over K_{j-1}, so that $[K_j:K_{j-1}] < \infty$. We have a tower of extensions, and consequently

$$[L:K] = [K_n:K_0] = [K_n:K_{n-1}][K_{n-1}:K_{n-2}] \ldots [K_1:K_0] < \infty.$$

Corollary 1 *If L:K is an extension and if α is an element of L which is algebraic over K, then $K(\alpha):K$ is algebraic.*

This is a special case of the next corollary.

Corollary 2 *Suppose that L:K is an extension and that $S \subset L$. If each $\alpha \in S$ is algebraic over K, then $K(S):K$ is algebraic.*

Proof. If $\beta \in K(S)$, there exist $\alpha_1, \ldots, \alpha_n$ in S such that $\beta \in K(\alpha_1, \ldots, \alpha_n)$. By the theorem, $K(\alpha_1, \ldots, \alpha_n):K$ is algebraic, and so β is algebraic over K.

The proof of this corollary shows that, even though an algebraic extension may be infinite, it is possible to deal with it by using arguments involving finite extensions. The same is true of the next result.

Theorem 4.7 *Suppose that $M:L$ and $L:K$ are algebraic extensions. Then $M:K$ is algebraic.*

Proof. Suppose that $\alpha \in M$, and that

$$m_\alpha = l_0 + l_1 x + \cdots + l_n x^n$$

is its minimal polynomial over L. (Since m_α is monic, $l_n = 1$.) Then α is algebraic over $K(l_0, \ldots, l_n)$ and so

$$[K(l_0, \ldots, l_n)(\alpha):K(l_0, \ldots, l_n)] = [K(l_0, \ldots, l_n, \alpha):K(l_0, \ldots, l_n)] < \infty$$

by Theorem 4.4. Also

$$[K(l_0, \ldots, l_n):K] < \infty$$

by Theorem 4.6, and so

$$[K(\alpha):K] \leqslant [K(l_0, \ldots, l_n, \alpha):K]$$
$$= [K(l_0, \ldots, l_n, \alpha):K(l_0, \ldots, l_n)][K(l_0, \ldots, l_n):K]$$
$$< \infty;$$

thus α is algebraic over K.

Exercises

4.7 Show that if $L:K$ is algebraic and K is countable then L is countable. Show that there exist real numbers which are transcendental over the rationals.

4.8 Suppose that $L:K$ is an extension, that α is an element of L which is transcendental over K, and that f is a non-constant element of $K[x]$. Show that $f(\alpha)$ is transcendental over K. Show that, if β is an element of L which satisfies $f(\beta) = \alpha$, then β is transcendental over K.

4.9 Suppose that a and b are complex numbers which are transcendental over $\mathbb{Q}$. Is a^b transcendental over $\mathbb{Q}$?

4.10 Suppose that $K(\alpha, \beta):K$ is an extension, that α is algebraic over K, but not in K, and that β is transcendental over K. Show that $K(\alpha, \beta):K$ is not simple.

4.5 Monomorphisms of algebraic extensions

The next result uses finiteness in a rather different way. If $L:K$ is an extension and $\tau:L \to L$ is a monomorphism with the property that $\tau(k)=k$ for each k in K, we say that τ *fixes* K.

Theorem 4.8 *Suppose that $L:K$ is algebraic and that $\tau:L \to L$ is a monomorphism which fixes K. Then τ maps L onto L.*

Proof. Certainly $\tau(0)=0$. Suppose that α is a non-zero element of L. Let m_α be its minimal polynomial over K. Let R be the set of roots of m_α in L. If $\beta \in R$,

$$m_\alpha(\tau(\beta)) = \tau(m_\alpha(\beta)) = \tau(0) = 0$$

so that τ maps R into R. Now τ is one–one and R is finite (see Exercise 3.15) and so τ must map R onto R. Thus there exists β in R such that $\tau(\beta)=\alpha$. As this holds for each α in L, τ must map L onto L.

A ring monomorphism of a field onto itself is called an *automorphism*.

Exercise

4.11 Show that the condition that $L:K$ is algebraic cannot be dropped from Theorem 4.8.

5

Tests for irreducibility

5.1 Introduction

Suppose that f is a polynomial in $K[x]$, where K is a field. Since $K[x]$ is a unique factorization domain, f can be expressed essentially uniquely as a product of irreducible polynomials. This raises the important practical problem: how do we recognize whether or not a given polynomial is irreducible?

There are many important cases when the field K which we consider is the field of fractions of a unique factorization domain R: this is so in the most important case of all, when the field is the field $\mathbb{Q}$ of rational numbers. In such a situation, Gauss' lemma (the corollary to Theorem 3.12) is particularly useful. Recall that Gauss' lemma implies that, if f is irreducible in $R[x]$, then f is irreducible in $K[x]$.

As an application (which we shall need in the next chapter) let us consider

$$f = x^3 - 3x - 1 \in \mathbb{Z}[x].$$

As f is a cubic, if it factorized in $\mathbb{Z}[x]$ it would have a linear factor, and this would have to be either $x - 1$ or $x + 1$. But $f(1) = -3$ and $f(-1) = 1$, and so f is irreducible in $\mathbb{Z}[x]$. By Gauss' lemma, f is irreducible in $\mathbb{Q}[x]$.

In order to show the importance of Gauss' lemma, let us sketch the proof of the following result, due to Kronecker:

Theorem 5.1 *There is an algorithm to express any element of $\mathbb{Z}[x]$ as a product of irreducible factors.*

An algorithm is a procedure which takes a finite number of steps; the number of steps depends upon the polynomial in question, but an upper bound can be given for it in each case.

Proof. Suppose that f has degree n. Let r be the greatest integer such that $2r \leqslant n$. If f is not irreducible, f must have a non-unit factor of degree less than or equal to r. We search for such a factor. Let $c_j = f(j)$, for $0 \leqslant j \leqslant r$. If $c_j = 0$ for some $0 \leqslant j \leqslant r$, then $x - j$ is a factor of f. Otherwise, if g is a factor of f in $\mathbb{Z}[x]$ then $g(j)$ must divide c_j for $0 \leqslant j \leqslant r$. Each c_j had finitely many divisors, and an algorithm exists to determine them. Suppose that $(d_0, \ldots, d_r)$ is such that d_j is a divisor of c_j for $0 \leqslant j \leqslant r$. There exists a unique polynomial g in $\mathbb{Q}[x]$ of degree at most r such that $g(j) = d_j$ for $0 \leqslant j \leqslant r$:

$$g = \sum_{j=0}^{r} d_j g_j,$$

where

$$g_j = \prod_{0 \leqslant k \leqslant r, k \neq j} \left(\frac{x-k}{j-k} \right).$$

We can now test (by further algorithms) whether $g \in \mathbb{Z}[x]$ and whether g divides f. As there are only finitely many $(r+1)$-tuples $(d_0, \ldots, d_r)$ to consider, this means that there is an algorithm to find a non-unit factor of f, if one exists. Repeated use of the algorithm leads to a factorization as a product of irreducible factors.

This result is of theoretical importance, but the procedure is too cumbersome to use in practice. Trial and error may enable us to factorize a polynomial, but will not establish that a polynomial is irreducible. It is therefore important to establish simple criteria which will ensure that a polynomial is irreducible. This is what we shall do in the present chapter.

Exercises

5.1 Write (an outline of) a computer programme to implement the algorithm of Theorem 5.1.

5.2 Suppose that

$$f = a_0 + \cdots + a_n x^n$$

is a polynomial in $\mathbb{Z}[x]$ of degree n, and that $\max_i |a_i| = K$. Obtain an upper bound, in terms of n and K, for the number of calculations required to determine whether or not f is irreducible.

5.3 Suppose that K is a field with finitely many elements. Show that there is an algorithm to express any element of $K[x]$ as a product of irreducible factors.

5.4 Suppose that K is a field and that f and g are relatively prime in $K[x]$. Show that $f - yg$ is irreducible in $K(y)[x]$.

5.5 Suppose that $K(\alpha):K$ is simple and that α is transcendental over K. Show that if $\beta \in K(\alpha)$ and $\beta \notin K$ then $K(\alpha):K(\beta)$ is finite and β is transcendental over K. Show that, if $\beta = f(\alpha)/g(\alpha)$, where f and g are relatively prime in $K[x]$, then

$$[K(\alpha):K(\beta)] = \max(\text{degree } f, \text{degree } g).$$

5.2 Eisenstein's criterion

Eisenstein's criterion is concerned with factorization in $R[x]$, where R is an integral domain.

Theorem 5.2 (Eisenstein's criterion) *Suppose that R is an integral domain, and that*

$$f = f_0 + f_1 x + \cdots + f_n x^n \in R[x]$$

has the property that $f_0, \ldots, f_n$ are relatively prime. Suppose that p is a prime in R, and that $p \mid f_i$ for $0 \leqslant i < n$, while p does not divide f_n and p^2 does not divide f_0. Then f is irreducible in $R[x]$.

Proof. Suppose that $f = gh$ where

$$g = g_0 + \cdots + g_r x^r$$

and

$$h = h_0 + \cdots + h_s x^s$$

are not units in R. If r were equal to 0 (so that $g = g_0$), it would follow that g_0 divides f_j for $0 \leqslant j \leqslant n$, so that g_0 would be a unit: this gives a contradiction, so that $r \geqslant 1$. Similarly $s \geqslant 1$. By hypothesis, p^2 does not divide $g_0 h_0$, so that p cannot divide both g_0 and h_0. Without loss of generality we may suppose that p does not divide h_0.

Now $g_r h_s = f_n$, so that, by hypothesis, p does not divide g_r. Let i be the least integer such that p does not divide g_i. Then $0 \leqslant i \leqslant r < n$, so that $p \mid f_i$; that is,

$$p \mid (h_0 g_i + h_1 g_{i-1} + \cdots + h_i g_0).$$

As $p \mid g_j$ for $j < i$, $p \mid h_0 g_i$. As p is a prime, $p \mid h_0$ or $p \mid g_i$, giving a contradiction.

As an example (which we shall need later on) let us observe that

$$f = x^5 - 4x + 2$$

is irreducible over $\mathbb{Z}[x]$, by Eisenstein's criterion (with $p = 2$), and so f is irreducible over $\mathbb{Q}[x]$, by Gauss' lemma.

Exercises

5.6 Suppose that R is an integral domain and that

$$f = f_0 + f_1 x + \cdots + f_n x^n \in R[x]$$

has the property that $f_0, \ldots, f_n$ are relatively prime. Suppose that p is a prime in R, and that $p \mid f_i$ for $1 \leq i \leq n$, while p does not divide f_0 and p^2 does not divide f_n. Show that f is irreducible in $R[x]$.

5.7 Show that if p is a prime number then $x^n - p$ is irreducible in $\mathbb{Q}[x]$.

5.8 Let A denote the field of real numbers which are algebraic over $\mathbb{Q}$. Show that $[A:\mathbb{Q}] = \infty$.

5.9 Show that the positive pth roots of 2 (as p varies over the primes) are linearly independent over $\mathbb{Q}$.

5.10 Show that $x^5 - 4x + 2$ and $x^4 - 4x + 2$ are irreducible over $\mathbb{Q}(i)$.

5.3   Other methods for establishing irreducibility

Even if Eisenstein's criterion cannot be applied directly, it is sometimes possible to apply it after making a suitable transformation. For example, if

$$f = x^4 + 4x^3 + 10x^2 + 12x + 7 \in \mathbb{Z}[x],$$

it is not possible to apply Eisenstein's criterion directly. If we write $y = x + 1$, we find that

$$f = y^4 + 4y^2 + 2.$$

As

$$g = x^4 + 4x^2 + 2$$

is irreducible in $\mathbb{Z}[x]$, by Eisenstein's criterion, f must be irreducible too. The problem of course is to find a suitable transformation: this is a matter of ingenuity and good fortune.

There is another technique which can sometimes prove useful when we are considering polynomials in $\mathbb{Z}[x]$. Suppose that p is a prime number: for each integer n, let $\bar{n}$ denote the image (mod p) of n under the quotient map from $\mathbb{Z}$ to $\mathbb{Z}_p$. This quotient map induces a ring homomorphism from $\mathbb{Z}[x]$ onto $\mathbb{Z}_p[x]$; if

$$f = a_0 + a_1 x + \cdots + a_n x^n \in \mathbb{Z}[x],$$

then

$$\bar{f} = \bar{a}_0 + \bar{a}_1 x + \cdots + \bar{a}_n x^n \in \mathbb{Z}_p[x].$$

Theorem 5.3 (Localization principle) *Suppose that*

$$f = a_0 + a_1 x + \cdots + a_n x^n \in \mathbb{Z}[x],$$

and that $a_0, \ldots, a_n$ are relatively prime. Suppose that p is a prime which does not divide a_n. If $\bar{f}$ is irreducible in $\mathbb{Z}_p[x]$, then f is irreducible in $\mathbb{Z}[x]$.
Proof. Suppose that f factors as $f = gh$, where g and h are not units. As in

the proof of Eisenstein's criterion, since $a_0, \ldots, a_n$ are relatively prime, degree $g \geq 1$ and degree $h \geq 1$; of course degree g + degree h = degree f.

As p does not divide a_n, degree $\bar{f}$ = degree f. As $\bar{f} = \bar{g}\bar{h}$, degree $\bar{f}$ = degree $\bar{g}$ + degree $\bar{h}$. As degree $\bar{g} \leq$ degree g, and degree $\bar{h} \leq$ degree h, we must have that degree $\bar{g}$ = degree $g \geq 1$ and degree $\bar{h}$ = degree $h \geq 1$. Thus $\bar{f} = \bar{g}\bar{h}$ is a non-trivial factorization of $\bar{f}$.

Notice that the localization principle can also be used to establish Eisenstein's criterion in $\mathbb{Z}[x]$. With the notation of Theorem 5.2, $\bar{f} = \bar{f}_n x^n$ (why?). Consequently, as $\bar{f} = \bar{g}\bar{h}$, $g_0 = 0$ (mod p) and $h_0 = 0$ (mod p) so that $f_0 = g_0 h_0 = 0$ (mod p^2), giving a contradiction.

To give another example of the use of localization, let us show that $f = x^n + px + p^2$ is irreducible in $\mathbb{Z}[x]$ (where p is a prime number). First observe that if α is a root of f in $\mathbb{Z}$, then $\alpha < 0$ and $\alpha = 0$ (mod p) so that $\alpha = -kp$ for some positive integer k. From this it follows that

$$(-k)^n p^{n-2} = k - 1,$$

which clearly has no solution. Thus, if $f = gh$ is a factorization in $\mathbb{Z}[x]$, degree $g \geq 2$ and degree $h \geq 2$. Arguing as before, $g_i = 0$ (mod p) for $i <$ degree g and $h_j = 0$ (mod p) for $j <$ degree h, so that

$$p = g_0 h_1 + g_1 h_0 = 0 \quad (\text{mod } p^2)$$

giving a contradiction.

Exercises

5.11 Show (by making the transformation $y = x - 1$) that if p is a prime number then $1 + x + \cdots + x^{p-1}$ is irreducible over $\mathbb{Q}$.

5.12 Let $\theta = 2\pi/7$. What is the minimal polynomial of $e^{i\theta}$ over $\mathbb{Q}$? What is the minimal polynomial of $2 \cos \theta$ over $\mathbb{Q}$?

6

Ruler-and-compass constructions

One of the problems that greatly exercised the Greek mathematicians and their successors was to find a method, using ruler and compass, to trisect a given angle. We shall show that this is not possible – it is not possible to trisect the angle $\pi/3$ – using the ideas of Chapter 4. It is remarkable that these ideas, which are really rather elementary, resolve the problem so decisively: an idea does not need to be complicated in order to be effective.

6.1   Constructible points

There are many constructions that one can carry out with ruler (straight-edge) and compasses alone. Many children, on first being given a pair of compasses, find out for themselves how to construct a regular hexagon (and so construct the angle $\pi/3$). I hope that you remember enough school geometry to know how to bisect an angle, to drop a perpendicular from a point to a line, to draw a line through a point parallel to a given line, and so divide an interval into a given rational ratio, using ruler and compasses alone.

Let us try and describe the situation in an accurate but informal way. We begin with two distinct points P_0 and P_1 in the plane. We take P_0 as origin, and take as our first axis the line through P_0 and P_1, and as our second axis the line through P_0 perpendicular to P_0P_1. We take P_0P_1 as our unit of distance. In this way, we can think of each point in the plane as an element (x, y) of $\mathbb{R} \times \mathbb{R}$. We call x and y the *coefficients* of the point. From what we have said in the preceding paragraph, we can certainly construct any point (r_1, r_2) with rational coefficients, using ruler and compasses alone. There are many other points that we can construct, too: let us describe more accurately what this means.

We shall say that a point P is *constructible* if there exists a finite sequence

$P_0, P_1, \ldots, P_n = P$ of points in the plane with the following property. Let
$$S_j = \{P_0, P_1, \ldots, P_j\}, \text{ for } 1 \leqslant j \leqslant n.$$
For each $2 \leqslant j \leqslant n$, P_j is *either*

   (i) the intersection of two distinct straight lines, each joining two points of S_{j-1}, *or*

  (ii) a point of intersection of a straight line joining two points of S_{j-1} and a circle with centre a point of S_{j-1} and radius the distance between two points of S_{j-1}, *or*

 (iii) a point of intersection of two distinct circles, each with centre a point of S_{j-1} and radius the distance between two points of S_{j-1}.

In case (iii), the centres *must* be different if the circles are to intersect: the radii may or may not be different.

We now wish to associate some fields to these geometric ideas. We do this in a very straightforward way: $\mathbb{R}:\mathbb{Q}$ is an extension; if $P = (x, y)$ is a constructible point, we consider the extension $\mathbb{Q}(x, y): \mathbb{Q}$ generated by x and y.

Theorem 6.1 *If $P = (x, y)$ is a constructible point, the extension $\mathbb{Q}(x, y):\mathbb{Q}$ is finite, and $[\mathbb{Q}(x, y):\mathbb{Q}] = 2^r$, for some non-negative integer r.*
Proof. Since P is constructible, there exists a sequence $P_0, P_1, \ldots, P_n = P$ of points which satisfies the requirements of the definitions. Let $P_j = (x_j, y_j)$, and for $1 \leqslant j \leqslant n$ let
$$F_j = \mathbb{Q}(x_1, y_1, x_2, y_2, \ldots, x_j, y_j).$$
Then $F_{j+1} = F_j(x_{j+1}, y_{j+1})$, for $1 \leqslant j < n$. We shall show that $[F_{j+1}:F_j] = 1$ or 2: then, by the tower law, $[F_n:F_1] = [F_n:\mathbb{Q}] = 2^s$ for some non-negative integer s. But $\mathbb{Q}(x, y) = \mathbb{Q}(x_n, y_n)$ is a subfield of F_n containing $\mathbb{Q}$, so that, by the tower law again,
$$[F_n:\mathbb{Q}(x, y)][\mathbb{Q}(x, y):\mathbb{Q}] = 2^s,$$
and so $[\mathbb{Q}(x, y):\mathbb{Q}] = 2^r$, for some non-negative integer r.

It remains to show that $[F_{j+1}:F_j] = 1$ or 2.

If (a_1, b_1) and (a_2, b_2) are two points in S_j, the equation of the line joining (a_1, b_1) and (a_2, b_2) is $(x - a_2)(b_1 - b_2) = (a_1 - a_2)(y - b_2)$, and therefore has the form
$$\lambda x + \mu y + \nu = 0,$$
where λ, μ and ν are elements of F_j. Similarly the equation of the circle, centre (a_1, b_1) and radius the distance between points (a_2, b_2) and (a_3, b_3) of S_j, is
$$(x - a_1)^2 + (y - b_1)^2 = (a_2 - a_3)^2 + (b_2 - b_3)^2,$$

and therefore has the form

$$x^2 + y^2 + 2gx + 2fy + c = 0,$$

where f, g and c are elements of F_j.

We are now in a position to consider the three cases that can arise.

Case (i). (x_{j+1}, y_{j+1}) is the intersection of two distinct straight lines, each joining two points of S_j. In this case (x_{j+1}, y_{j+1}) is the solution of two simultaneous equations

$$\lambda_1 x + \mu_1 y + \nu_1 = 0,$$

$$\lambda_2 x + \mu_2 y + \nu_2 = 0$$

with coefficients in F_j. Solving these, we find that x_{j+1} and y_{j+1} are in F_j, so that $F_{j+1} = F_j$ and $[F_{j+1} : F_j] = 1$.

Case (ii). (x_{j+1}, y_{j+1}) is a point of intersection of an appropriate straight line and circle. In this case (x_{j+1}, y_{j+1}) satisfies equations

$$\lambda x + \mu y + \nu = 0,$$

$$x^2 + y^2 + 2gx + 2fy + c = 0$$

with coefficients in F_j. Suppose that $\lambda \neq 0$. We can then eliminate x, and obtain a monic quadratic equation in y. If this factors over F_j as

$$(y - \alpha)(y - \beta) = 0$$

then $y_{j+1} = \alpha$ or β, so that $y_{j+1} \in F_j$; substituting in the linear equation, $x_{j+1} \in F_j$, so that $F_{j+1} = F_j$, and $[F_{j+1} : F_j] = 1$. If the quadratic is irreducible, it must be the minimal polynomial for y_{j+1}: thus, by Theorem 4.4, $[F_j(y_{j+1}) : F_j] = 2$. As $x_{j+1} = -\lambda^{-1}(\mu y_{j+1} + \nu)$, $x_{j+1} \in F_j(y_{j+1})$ and so $F_{j+1} = F_j(x_{j+1}, y_{j+1}) = F_j(y_{j+1})$. If $\lambda = 0$, then $\mu \neq 0$, and we can repeat the argument, interchanging the roles of x_{j+1} and y_{j+1}.

Case (iii). (x_{j+1}, y_{j+1}) is a point of intersection of two suitable circles. In this case (x_{j+1}, y_{j+1}) satisfies equations

$$x^2 + y^2 + 2g_1 x + 2f_1 y + c_1 = 0,$$

$$x^2 + y^2 + 2g_2 x + 2f_2 y + c_2 = 0$$

with coefficients in F_j. Subtracting, (x_{j+1}, y_{j+1}) satisfies the equation

$$2(g_1 - g_2)x + 2(f_1 - f_2)y + (c_1 - c_2) = 0.$$

We cannot have $g_1 = g_2$ and $f_1 = f_2$, for then the circles would be concentric, and would not intersect. Thus this case reduces to the previous one.

Although the proof of this theorem may appear to be rather lengthy, you should note that almost all the field theory appears in the first paragraph: the rest is coordinate geometry of a particularly simple kind.

Exercises

6.1 Describe how the constructions mentioned in the first paragraph of Section 6.1 are made.

6.2 Suppose that the point $(l, 0)$ is constructible (where $l > 0$). Show how to construct the points $(\sqrt{l}, 0)$ and $(l^2, 0)$.

6.3 Construct a regular pentagon.

6.4 Suppose that $P = (x, y)$ is a constructible point. Let α and β be elements of $\mathbb{Q}(x, y)$. Show that (α, β) is a constructible point.

6.2 The angle π/3 cannot be trisected

We have observed that, using ruler and compasses, we can construct the angle $\pi/3$. We shall now show that this angle cannot be trisected, using ruler and compasses alone, in the way that we have described. Let us write α for $\pi/9$. If we could trisect $\pi/3$, there would be a constructible point P, other than P_0, on the line given by

$$x \sin \alpha = y \cos \alpha;$$

intersecting the line PP_0 with the circle with centre P_0 and radius P_0P_1, it would follow that $(\cos \alpha, \sin \alpha)$ would be constructible. As

$$[\mathbb{Q}(\cos \alpha, \sin \alpha) : \mathbb{Q}] = [\mathbb{Q}(\cos \alpha, \sin \alpha) : \mathbb{Q}(\cos \alpha)][\mathbb{Q}(\cos \alpha) : \mathbb{Q}]$$

it would follow from the tower law and Theorem 6.1 that $[\mathbb{Q}(\cos \alpha) : \mathbb{Q}] = 2^t$, for some non-negative t. Now recall that

$$\cos 3\theta = 4(\cos \theta)^3 - 3 \cos \theta$$

and that $\cos(\pi/3) = \frac{1}{2}$, so that

$$4(\cos \alpha)^3 - 3 \cos \alpha - \tfrac{1}{2} = 0.$$

Let $\sigma = 2 \cos \alpha$. Then

$$\sigma^3 - 3\sigma - 1 = 0.$$

As we have seen in Section 5.1, $x^3 - 3x - 1$ is irreducible over $\mathbb{Q}$, and is therefore the minimal polynomial for σ over $\mathbb{Q}$. Thus, by Theorem 4.4, $[\mathbb{Q}(\sigma) : \mathbb{Q}] = 3$. But $\mathbb{Q}(\sigma) = \mathbb{Q}(\cos \alpha)$, and so we have the required contradiction.

Exercises

6.5 Show that the point $(2^{1/3}, 0)$ is not constructible (impossibility of 'duplicating the cube').

6.6 Show that it is not possible to construct (a) a regular nonagon or (b) a regular heptagon, using ruler and compasses.

6.3   Concluding remarks

Theorem 6.1 provides a necessary condition which a constructible point must satisfy. Is it a sufficient condition for a point to be constructible and, if not, what is a sufficient condition? These are much more difficult questions than those which we have answered in this chapter. Bear them in mind as the theory develops.

Exercise

6.7 (a) Suppose that x and y are real numbers such that $[\mathbb{Q}(x, y):\mathbb{Q}] = 2$. Show that (x, y) is constructible.

(b) Suppose that x and y are real numbers, and that $\mathbb{Q} = F_0 \subseteq F_1 \subseteq \cdots \subseteq F_r = \mathbb{Q}(x, y)$ is an increasing sequence of fields such that $[F_{j+1}:F_j] = 2$ for $0 \leqslant j < r$. Show, by induction on r, that (x, y) is constructible.

7

Splitting fields

Suppose first that $f \in \mathbb{Q}[x]$. As we have seen, f may be irreducible; if not, we can factorize f in an essentially unique way into irreducible factors. This is as far as factorization can go in $\mathbb{Q}[x]$.

On the other hand, $\mathbb{Q}$ is a subfield of $\mathbb{C}$ and we can consider f as an element of $\mathbb{C}[x]$. Now the field $\mathbb{C}$ has the remarkable property that any non-unit element of $\mathbb{C}[x]$ can be written as a product of linear factors. This is, of course, an immediate consequence of the fact that any non-constant polynomial p in $\mathbb{C}[x]$ has a root in $\mathbb{C}$. This fact (the 'fundamental theorem of algebra') is usually proved by complex function theory: if p had no root, $1/p$ would be a non-constant bounded analytic function on $\mathbb{C}$, contradicting Liouville's theorem. (You may feel that there is too much analysis in this. Some analysis is certainly needed, since the real field $\mathbb{R}$ is an analytic construction. Be patient, and be sure to tackle Exercise 11.11 in due course.)

If we consider f as an element of $\mathbb{C}[x]$, then, we can write

$$f = \lambda(x - \alpha_1) \ldots (x - \alpha_n),$$

where λ is a rational number and $\alpha_1, \ldots, \alpha_n$ are complex numbers. Each α_i is algebraic over $\mathbb{Q}$, since $f(\alpha_i) = 0$. Thus, if $L = \mathbb{Q}(\alpha_1, \ldots, \alpha_n)$, L is algebraic over $\mathbb{Q}$ and L is finitely generated over $\mathbb{Q}$, and so $[L:\mathbb{Q}] < \infty$ (Theorem 4.6). Further f factorizes into linear factors over L. As far as f is concerned, then, L is large enough for our purposes.

The above argument works because of the special properties of the complex field $\mathbb{C}$. Our aim in this chapter is to show how, starting with an element f of $K[x]$, where K is an arbitrary field, we can construct a finite extension $L:K$ such that f factorizes into linear factors over L.

Exercise

7.1 Suppose that f is an irreducible polynomial in $\mathbb{R}[x]$. Show that degree $f \leqslant 2$.

7.1   Splitting fields

Suppose that K is a field, that $f \in K[x]$ and that $L:K$ is an extension. We say that f *splits over L* if we can write

$$f = \lambda(x - \alpha_1) \ldots (x - \alpha_n)$$

where $\alpha_1, \ldots, \alpha_n$ are in L and $\lambda \in K$.

We say that $L:K$ is a *splitting field extension for f over K* (or, when it is clear what K is, that L is a *splitting field* for f) if, first, f splits over L and, secondly, there is no proper subfield L' of L containing K such that f splits over L'. This last condition ensures that the extension $L:K$ is an economical one for f.

If we can find an extension over which f splits, we can find a splitting field:

Theorem 7.1 *Suppose that $L:K$ is an extension and that $f \in K[x]$ splits over L as*

$$f = \lambda(x - \alpha_1) \ldots (x - \alpha_n).$$

Then $K(\alpha_1, \ldots, \alpha_n)$ is a splitting field for f.
Proof. f certainly splits over $K(\alpha_1, \ldots, \alpha_n)$. Suppose that $K(\alpha_1, \ldots, \alpha_n) \supseteq K' \supseteq K$ and that f splits over K':

$$f = \lambda'(x - \alpha_1') \ldots (x - \alpha_n').$$

As factorization in $L[x]$ is essentially unique, for each i we have $\alpha_i = \alpha_j'$, for some j, and so $\alpha_i \in K'$. Consequently $K' \supseteq K(\alpha_1, \ldots, \alpha_n)$ and so K' is not a proper subfield of $K(\alpha_1, \ldots, \alpha_n)$.

Corollary *If $L:K$ is a splitting field extension for $f \in K[x]$ then $L:K$ is a finite algebraic extension.*

How can we construct splitting fields? The key step is the adjunction of a root of an irreducible polynomial.

Theorem 7.2 *Suppose that $f \in K[x]$ is irreducible of degree n. Then there is a simple algebraic extension $K(\alpha):K$ such that $[K(\alpha):K] = n$ and $f(\alpha) = 0$.*
Proof. We must construct $K(\alpha)$ intrinsically, starting from K and f. Let $j: K \to K[x]$ be the natural monomorphism, let $L = K[x]/(f)$, and let $q: K[x] \to L$ be the quotient map. Since f is irreducible, L is a field (by the corollary to Theorem 3.16). Let $i = qj:i$ is a monomorphism of the field K into the field L

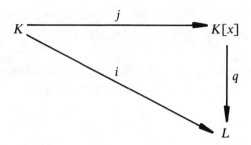

so that (i, K, L) is an extension (remember the original definition). Now let $\alpha = q(x) = x + (f)$. As x generates $K[x]$ over K, $L = K(\alpha)$. Also, since q is a ring homomorphism,

$$f(\alpha) = f(q(x)) = q(f) = 0.$$

Thus α is algebraic over K. Bearing in mind that f is irreducible over K, we see that f must be a scalar multiple of the minimal polynomial m_α of α over K. Thus $[L:K] = n$, by Theorem 4.4.

Note that, although f is irreducible over K, it is not irreducible over $K(\alpha)$: it has a linear factor $x - \alpha$. Factorization is under way, and we can now proceed inductively.

Theorem 7.3 Suppose that $f \in K[x]$. Then there exists a splitting field extension $L:K$ for f, with $[L:K] \leqslant n!$.
Proof. We prove this by induction on $n =$ degree f. Of course, if degree $f \leqslant 1$, there is nothing to prove. Suppose that the result holds for any polynomial of degree less than n, over any field K. Suppose that degree $f = n$. We consider two cases.
Case 1. f is not irreducible over K. We can write $f = gh$, where degree $g = s < n$ and degree $h = t < n$. By the inductive hypothesis there is a splitting field $L:K$ for g, with $[L:K] \leqslant s!$. We can write

$$g = \lambda(x - \alpha_1) \ldots (x - \alpha_s)$$

with $\alpha_i \in L$ and $\lambda \in K$. Note that $L = K(\alpha_1, \ldots, \alpha_s)$.

We can now consider h as an element of $L[x]$; by the inductive hypothesis again, there is a splitting field $M:L$ for h, with $[M:L] \leqslant t!$ We can write

$$h = \mu(x - \beta_1) \ldots (x - \beta_t)$$

with $\beta_i \in M$ and $\mu \in L$. Note that $M = L(\beta_1, \ldots, \beta_t) = K(\alpha_1, \ldots, \alpha_s, \beta_1, \ldots, \beta_t)$; as $\lambda\mu$ is the coefficient of x^n in f, $\lambda\mu \in K$. Thus $M:K$ is a splitting field extension for f. Further,

$$[M:K] = [M:L][L:K] \leqslant t! s! \leqslant (s+t)! = n!$$

Case 2. f is irreducible over K. Then by Theorem 7.2 there exists a simple algebraic extension $K(\alpha):K$, with $[K(\alpha):K] = n$, such that, over $K(\alpha)$,

$$f = (x - \alpha)h$$

where $h \in K(\alpha)[x]$, and degree $h = n - 1$. By the inductive hypothesis, there exists a splitting field extension $L:K(\alpha)$ for h, with $[L:K(\alpha)] \le (n-1)!$. We can write

$$h = \mu(x - \beta_1) \ldots (x - \beta_{n-1})$$

with $\beta_i \in L$, $\mu \in K(\alpha)$. Note that $L = K(\alpha)(\beta_1, \ldots, \beta_{n-1}) = K(\alpha, \beta_1, \ldots, \beta_{n-1})$. Then

$$f = \mu(x - \alpha)(x - \beta_1) \ldots (x - \beta_{n-1});$$

again, μ is the coefficient of x^n in f, so that $\mu \in K$ and f splits over L. Thus

$$L:K = K(\alpha, \beta_1, \ldots, \beta_{n-1}):K$$

is a splitting field extension for f. Finally

$$[L:K] = [L:K(\alpha)][K(\alpha):K] \le (n-1)! n = n!$$

Observe that the proof of Theorem 7.3 is largely a matter of induction; the field theory occurs in Theorem 7.2.

Nevertheless, Theorem 7.3 is a major achievement: we can now produce a splitting field for *any* polynomial over *any* field. Notice that there can be some freedom of action in Theorem 7.3 (in the way we consider factors in the case where f is not irreducible); there may also be other ways to produce splitting fields: can these be essentially different? We shall answer this important question in the next section.

Exercises

7.2 Show that $f = x^3 - x + 1$ is irreducible in $\mathbb{Z}_3[x]$. Show that if ζ is a root of f in a splitting field extension, then $\zeta + 1$ and $\zeta - 1$ are also roots. Construct a splitting field extension, and write out its multiplication table.

7.3 Suppose that K is a field over which $x^n - 1$ splits, and suppose that $K(t):K$ is a simple transcendental extension. Show that $x^n - t$ is irreducible in $K(t)[x]$ (Exercise 5.4). Construct a splitting field extension for $x^n - t$ by considering another simple transcendental extension $K(s):K$ and a monomorphism $i: K(t) \to K(s)$ which fixes K and sends t to s^n.

7.2 The extension of monomorphisms

In this section, we shall show that a splitting field extension of a polynomial is essentially unique. In the process, we shall prove some of the most important results of the theory. As the theory develops in the succeeding chapters, it will, I hope, become clear why these results are so important. Two more remarks are in order. First, algebra is not just the study of sets with some algebraic structure, but the study of such sets and of mappings between them which respect the structure: in this section we begin to consider such mappings. Secondly, although the results are important, the proofs are natural and easy: the relationship between 'difficulty' and 'importance' is a curious one.

Let us recall that if i is a ring homomorphism from a field K into a field L then i is necessarily a monomorphism, so that i is an isomorphism of K onto $i(K)$. Further if

$$f = a_0 + a_1 x + \cdots + a_n x^n \in K[x]$$

then $i(f) = i(a_0) + i(a_1)x + \cdots + i(a_n)x^n \in i(K)[x] \subseteq L[x]$; thus i extends to a monomorphism (which we again denote by i) from $K[x]$ into $L[x]$, and i is an isomorphism of $K[x]$ onto $i(K)[x]$.

We begin by considering simple algebraic extensions.

Theorem 7.4 *Suppose that $K(\alpha):K$ is a simple extension and that α is algebraic over K, with minimal polynomial m_α. Suppose that i is a monomorphism from K into a field L and that $\beta \in L$. Then a necessary and sufficient condition for there to be a monomorphism j from $K(\alpha)$ to L with $j(\alpha) = \beta$ and $j|_K = i$ is that $i(m_\alpha)(\beta) = 0$. If the condition is satisfied then j is unique.*

Proof. Necessity. This is rather trivial. If j exists then

$$i(m_\alpha)(\beta) = j(m_\alpha)(j(\alpha)) = j(m_\alpha(\alpha)) = j(0) = 0.$$

Sufficiency. Suppose that the condition is satisfied. Let $K' = i(K)$. Then $i: K \to K'$ is an isomorphism, which extends to an isomorphism $i: K[x] \to K'[x]$. As $i(m_\alpha)(\beta) = 0$, β is algebraic over K'. We now use the evaluation maps to construct the following diagram.

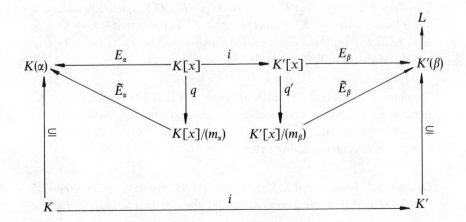

Here E_α and E_β are the evaluation maps, q and q' are quotient maps and $\tilde{E}_\alpha$ and $\tilde{E}_\beta$ are isomorphisms.

Now $i(m_\alpha)$ is monic, and is irreducible over K' (since i is an isomorphism of $K[x]$ onto $K'[x]$ which sends K to K') and $i(m_\alpha)(\beta)=0$, by hypothesis. Consequently $m_\beta = i(m_\alpha)$, and so (m_α) is the kernel of $q'i$. Thus by Theorem 3.1 there exists an isomorphism

$$\tilde{i}: K[x]/(m_\alpha) \to K'[x]/(m_\beta)$$

such that $q'i = \tilde{i}q$. Now let

$$j = \tilde{E}_\beta \tilde{i}(\tilde{E}_\alpha)^{-1}.$$

j is an isomorphism of $K(\alpha)$ onto $K'(\beta)$, and so it is a monomorphism of $K(\alpha)$ into L. Also

$$j(\alpha) = \tilde{E}_\beta \tilde{i}(\tilde{E}_\alpha)^{-1}(\alpha) = \tilde{E}_\beta \tilde{i}q(x)$$
$$= \tilde{E}_\beta q'i(x) = E_\beta(i(x)) = \beta$$

and if $k \in K$

$$j(k) = \tilde{E}_\beta \tilde{i}(\tilde{E}_\alpha)^{-1}(k) = \tilde{E}_\beta \tilde{i}q(k)$$
$$= \tilde{E}_\beta q'i(k) = E_\beta(i(k)) = i(k),$$

so that j has the properties that we are looking for.

Finally, if j' is another monomorphism of $K(\alpha)$ with the required properties, then the set

$$F = \{\gamma : j(\gamma) = j'(\gamma)\}$$

is a subfield of $K(\alpha)$. It contains K and α, and so $F = K(\alpha)$ and j is unique.

This theorem can be proved more quickly: it is not really necessary to show that $i(m_\alpha) = m_\beta$. But this proof shows how rigidly j is determined: we have built a strong bridge between $K(\alpha)$ and $K'(\beta)$.

Inspection of the diagram and the proof, gives the following corollary.

Corollary 1 *Suppose that $K(\alpha):K$ and $K'(\alpha'):K'$ are simple extensions, and that α is algebraic over K, α' algebraic over K'. Suppose that $i: K \to K'$ is an isomorphism. Then there exists an isomorphism $j: K(\alpha) \to K'(\alpha')$ with $j(\alpha) = \alpha'$ and $j|_K = i$ if and only if $i(m_\alpha) = m_{\alpha'}$. If so, j is unique.*

Corollary 2 *Suppose that $K(\alpha):K$ is simple and that α is algebraic over K. Suppose that $i: K \to L$ is a monomorphism, and that $i(m_\alpha)$ has r distinct roots in L. Then there are exactly r distinct monomorphisms $j: K(\alpha) \to L$ with $j|_K = i$.*

We now consider splitting fields.

Theorem 7.5 *Suppose that $\Sigma:K$ is a splitting field extension for a polynomial f in $K[x]$ and that i is a monomorphism from K into a field L. Then a* *necessary and sufficient condition for there to be a monomorphism j from* Σ *into L with* $j|_K = i$ *is that* $i(f)$ *splits over L.*

Proof. Necessity. As f splits over Σ, we can write

$$f = \lambda(x - \alpha_1) \ldots (x - \alpha_n),$$

with $\lambda \in K, \alpha_1, \ldots, \alpha_n \in \Sigma$. Then

$$i(f) = j(f) = i(\lambda)(x - j(\alpha_1)) \ldots (x - j(\alpha_n))$$

so that $i(f)$ splits over L.

Sufficiency. Once again we argue by induction on degree $f = n$. The result is true when $n = 1$, for then $\Sigma = K$, and we take $j = i$. Suppose that the result holds for any splitting field extension $\Sigma' : K'$ for any polynomial of degree less than n over any field K', and for any monomorphism i' from K' into L. Suppose that degree $f = n$, and that $i(f)$ splits over L.

As $\Sigma : K$ is a splitting field extension for f over K, we can write

$$f = \lambda(x - \alpha_1) \ldots (x - \alpha_n),$$

with $\alpha_i \in \Sigma$ and $\lambda \in K$. α_1 is algebraic over K; let m be its minimal polynomial over K. Then $f = mg$, and m is irreducible over K. By relabelling $\alpha_1, \ldots, \alpha_n$ if necessary, we can suppose that

$$m = (x - \alpha_1)(x - \alpha_2) \ldots (x - \alpha_r).$$

Now $i(f) = i(m)i(g)$; as $i(f)$ splits over L, $i(m)$ must split over L too. We can write

$$i(m) = (x - \beta_1) \ldots (x - \beta_r).$$

We are now in a position to apply Theorem 7.4: $K(\alpha_1) : K$ is a simple algebraic extension, and α_1 has minimal polynomial m. Also $i(m)(\beta_1) = 0$. There therefore exists a unique monomorphism j_1 from $K(\alpha_1)$ to L such that $j_1(\alpha_1) = \beta_1$ and $j_1|_K = i$:

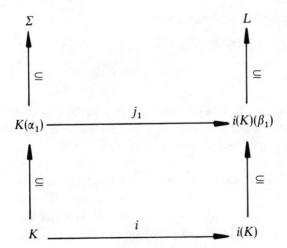

We now consider f as an element of $K(\alpha_1)[x]$. We can write $f = (x - \alpha_1)h$, where $h \in K(\alpha_1)[x]$, and h splits over Σ:

$$h = \lambda(x - \alpha_2) \ldots (x - \alpha_n).$$

Also $\Sigma = K(\alpha_1)(\alpha_2, \ldots, \alpha_n)$, and so Σ is a splitting field for h over $K(\alpha_1)$. As degree $h = n - 1$, we can apply the inductive hypothesis: there exists a monomorphism j from Σ to L such that $j|_{K(\alpha_1)} = j_1$. This completes the proof.

Before we establish some corollaries, let us make three remarks. First, like Theorem 7.4, this is an *extension* theorem: we extend the mapping i. Secondly, unlike Theorem 7.4, the extension need not be unique: we could map α_1 to any of $\beta_1, \ldots, \beta_r$. Thirdly, although the extension need not be unique, there are obviously some limitations on the number of extensions that there can be. This is a topic to which we shall pay much attention later on.

Corollary 1 *Suppose that* $i: K \to K'$ *is an isomorphism and that* $f \in K[x]$. *Suppose that* $\Sigma : K$ *is a splitting field extension for* f, $\Sigma' : K'$ *a splitting field extension for* $i(f)$. *Then there exists an isomorphism* $j: \Sigma \to \Sigma'$ *such that* $j|_K = i$.

Proof. If we apply the theorem to the mapping i, considered as a monomorphism from K to Σ', it follows that there exists a monomorphism j from Σ to Σ' which extends i. We can write

$$f = \lambda(x - \alpha_1) \ldots (x - \alpha_n),$$

with $\alpha_1, \ldots, \alpha_n$ in Σ and λ in K. Then

$$j(f) = i(\lambda)(x - j(\alpha_1)) \ldots (x - j(\alpha_n)),$$

so that, using Theorem 7.1, it follows that

$$\Sigma' = K'(j(\alpha_1), \ldots, j(\alpha_n)) \subseteq j(\Sigma),$$

and j is onto.

Corollary 2 *Suppose that* $f \in K[x]$ *is irreducible and that* $\Sigma : K$ *is a splitting field extension for* f. *If* α *and* β *are roots of* f *in* Σ, *there is an automorphism* $\sigma : \Sigma \to \Sigma$ *such that* $\sigma(\alpha) = \beta$ *and* σ *fixes* K.

Proof. We may suppose that f is monic: then f is the minimal polynomial for α and β over K. By Corollary 1 of Theorem 7.4, there is an isomorphism $\tau : K(\alpha) \to K(\beta)$ with $\tau(\alpha) = \beta$ and $\tau(k) = k$ for $k \in K$. Now $\Sigma : K(\alpha)$ is a splitting field extension for f over $K(\alpha)$, and $\Sigma : K(\beta)$ is a splitting field extension for f over $K(\beta)$. The result now follows from Corollary 1.

Exercise

7.4 The complex numbers $i\sqrt{3}$ and $1+i\sqrt{3}$ are roots of the quartic $f = x^4 - 2x^3 + 7x^2 - 6x + 12$. Does there exist an automorphism σ of the splitting field extension for f over $\mathbb{Q}$ with $\sigma(i\sqrt{3}) = 1 + i\sqrt{3}$?

7.3   Some examples

We now consider some examples of splitting fields. First let us consider polynomials in $\mathbb{Q}[x]$. If $f \in \mathbb{Q}[x]$ then, as we saw at the beginning of this chapter, f splits over $\mathbb{C}[x]$, and we can, and usually shall, consider the splitting field of f as a subfield of $\mathbb{C}$. Alternatively, we can make the constructions of Theorem 7.2 and 7.3. Corollary 1 to Theorem 7.5 then says that the splitting field that we obtain is essentially the same.

Example 1 $f = x^p - 2$ in $\mathbb{Q}[x]$ (with p a prime).

f is irreducible, by Eisenstein's criterion, and there is one real positive root $2^{1/p}$. f is the minimal polynomial of $2^{1/p}$, so that $[\mathbb{Q}(2^{1/p}):\mathbb{Q}] = p$. If α is any root of f in $\mathbb{C}$, then $(\alpha/2^{1/p})^p = \alpha^p/2 = 1$, so that $\alpha = 2^{1/p}\omega$, where ω is a root of $x^p - 1$. $x^p - 1$ is not irreducible, as

$$x^p - 1 = (x - 1)(x^{p-1} + x^{p-2} + \cdots + 1).$$

Now $x^{p-1} + x^{p-2} + \cdots + 1$ is irreducible over $\mathbb{Q}$ (Exercise 5.11), so that if ω is any root of $x^p - 1$ other than 1 then $[\mathbb{Q}(\omega):\mathbb{Q}] = p - 1$. The map $n \to \omega^n$ is a homomorphism of $\mathbb{Z}$ into the multiplicative group $\mathbb{C}^*$, with kernel $p\mathbb{Z}$, and so the complex numbers $1, \omega, \ldots, \omega^{p-1}$ must be distinct. They are all roots of $x^p - 1$, so that

$$x^p - 1 = (x - 1)(x - \omega) \ldots (x - \omega^{p-1})$$

and $\mathbb{Q}(\omega):\mathbb{Q}$ is a splitting field extension for $x^p - 1$.

Now our original polynomial f splits over $\mathbb{Q}(\omega, 2^{1/p})$ since it has roots

$$2^{1/p}, \omega 2^{1/p}, \ldots, \omega^{p-1} 2^{1/p}.$$

Further any splitting field must contain $2^{1/p}$, and must also contain $\omega = \omega 2^{1/p}/2^{1/p}$. Thus $\mathbb{Q}(\omega, 2^{1/p}):\mathbb{Q}$ is the splitting field extension for f.

What is $[\mathbb{Q}(\omega, 2^{1/p}):\mathbb{Q}]$? In order to answer this, consider this diagram.

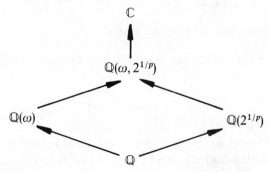

Here (and later, when we consider similar diagrams) rising arrows represent inclusion mappings.

By the tower law,

$$[\mathbb{Q}(2^{1/p}):\mathbb{Q}]\big|[\mathbb{Q}(\omega,2^{1/p}):\mathbb{Q}] \quad \text{and} \quad [\mathbb{Q}(\omega):\mathbb{Q}]\big|[\mathbb{Q}(\omega,2^{1/p}):\mathbb{Q}].$$

As $[\mathbb{Q}(2^{1/p}):\mathbb{Q}]=p$ and $[\mathbb{Q}(\omega):\mathbb{Q}]=p-1$, this means that $[\mathbb{Q}(\omega,2^{1/p}):\mathbb{Q}]\geqslant p(p-1)$. On the other hand, if m is the minimal polynomial of $2^{1/p}$ over $\mathbb{Q}(\omega)$, m divides x^p-2 in $\mathbb{Q}(\omega)[x]$, and so

$$\text{degree } m=[\mathbb{Q}(\omega,2^{1/p}):\mathbb{Q}(\omega)]\leqslant p.$$

Thus, by the tower law,

$$[\mathbb{Q}(\omega,2^{1/p}):\mathbb{Q}]=[\mathbb{Q}(\omega,2^{1/p}):\mathbb{Q}(\omega)][\mathbb{Q}(\omega):\mathbb{Q}]$$

$$\leqslant p(p-1),$$

and so $[\mathbb{Q}(\omega,2^{1/p}):\mathbb{Q}]=p(p-1)$.

This implies that degree $m=p$, and so x^p-2 is irreducible over $\mathbb{Q}(\omega)$.

This example has many important features. It is perhaps a bit more complicated than one might imagine. Notice that the pth roots of unity (the roots of x^p-1) played an important rôle. Notice also that we picked one of them (other than 1): had we picked another, the result would have been the same. (Can you formalize this, using Corollary 2 of Theorem 7.5?) Notice also that the argument could have been simplified by appealing to Exercise 4.2.

Example 2 $f=x^6-1$ in $\mathbb{Q}[x]$.
f factorizes as

$$f=(x-1)(x^2+x+1)(x+1)(x^2-x+1)$$

If ω is a root of x^2+x+1 then

$$f=(x-1)(x-\omega)(x-\omega^2)(x+1)(x+\omega)(x+\omega^2)$$

Thus $\mathbb{Q}(\omega):\mathbb{Q}$ is a splitting field extension for f and $[\mathbb{Q}(\omega):\mathbb{Q}]=2$.

Example 3 $f=x^6+1$ in $\mathbb{Q}[x]$.
The roots of f in $\mathbb{C}$ are $i, i\omega, i\omega^2, -i, -i\omega, -i\omega^2$. Thus, arguing as before, $\mathbb{Q}(i,\omega):\mathbb{Q}$ is the splitting field extension for f, and we have this diagram.

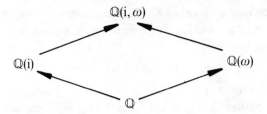

Now we can take $\omega = -\frac{1}{2} + \sqrt{3}\,i/2$, so that $\omega \notin \mathbb{Q}(i)$ (which consists of all complex numbers of the form $r + is$, with r and s in $\mathbb{Q}$). Thus $\mathbb{Q}(i) \neq \mathbb{Q}(\omega)$ and both $\mathbb{Q}(i)$ and $\mathbb{Q}(\omega)$ are proper subfields of $\mathbb{Q}(i, \omega)$. It is now easy to conclude that $[\mathbb{Q}(i, \omega):\mathbb{Q}] = 4$.

We now consider examples over more general fields. (To what extent do we use the fact that we are considering polynomials over the rationals in Examples 1 to 3?)

Example 4 $f = x^2 + ax + b$ in $K[x]$.

We would like to 'complete the square' and write

$$f = \left(x + \frac{a}{2}\right)^2 - \frac{a^2 - 4b}{4}.$$

As we shall see, this is not possible if char $K = 2$. Let us suppose that char $K \neq 2$. Then $2 = 1 + 1$ is a non-zero element of K, as are $4, \frac{1}{2}$ and $\frac{1}{4}$. (Thus in $\mathbb{Z}_3$, $2 = \frac{1}{2}$ and $1 = 4 = \frac{1}{4}$.)

In this case we can complete the square. We therefore consider the polynomial

$$g = x^2 - \mu, \text{ where } \mu = (a^2 - 4b)/4.$$

g splits over K if and only if there is an element v of K such that $v^2 = \mu$; in this case $g = (x - v)(x + v)$ and

$$f = \left(x + \frac{a}{2} - v\right)\left(x + \frac{a}{2} + v\right)$$

splits over K. If g is irreducible, there is a splitting field extension $L:K$ of degree 2. In this case $g = (x - v)(x + v)$, where v is an element of L not in K. Thus $L = K(v)$. Since once again

$$f = \left(x + \frac{a}{2} - v\right)\left(x + \frac{a}{2} + v\right)$$

$K(v):K$ is a splitting field extension for f.

To sum up: either μ has a square root in K, in which case f splits over K, or we obtain a splitting field by adjoining a square root of μ.

Note that, as a special case, $\mathbb{C} = \mathbb{R}(i)$ is a splitting field for $x^2 + 1$ over $\mathbb{R}$.

There remains the case where char $K = 2$. Let us restrict attention to the case where $K = \mathbb{Z}_2$. As $\mathbb{Z}_2$ has only two elements, there are only four monic quadratic polynomials:

$$x^2 = x \cdot x; \quad x^2 + x = x(x+1); \quad x^2 + 1 = (x+1)^2; \quad x^2 + x + 1.$$

The first three split over $\mathbb{Z}_2$, but $f = x^2 + x + 1$ is irreducible, since $f(0) = f(1) = 1$. By Theorem 7.3, there exists a splitting field extension $L : \mathbb{Z}_2$ for f and $[L:\mathbb{Z}_2] = 2$. Thus L has four elements, 0, 1, α and β say. The element $\alpha + 1$ is not in $\mathbb{Z}_2$, and so $\beta = \alpha + 1$. Thus $\alpha + \beta = 1$. The element $\alpha\beta$ is not zero (since L is a field) and cannot be α or β: thus $\alpha\beta = 1$. Consequently

$$f = x^2 + x + 1 = (x - \alpha)(x - \beta)$$

and α and β are the roots of f. Thus

$$\alpha^2 = \alpha + 1 = \beta \quad \text{and} \quad \beta^2 = \beta + 1 = \alpha$$

so L is not obtained by adjoining a square root.

Note though that $x^3 - 1 = (x - 1)(x^2 + x + 1)$ so that $L : \mathbb{Z}_2$ is a splitting field for $x^3 - 1$; $\alpha^3 = \beta^3 = 1$, and so L is obtained by adjoining cube roots.

Similar phenomena occur whenever we deal with finite fields. We shall consider these in more detail in Chapter 12.

Exercises

7.5 Suppose that $M:L$ and $L:K$ are extensions, and that $\alpha \in M$ is algebraic over K. Does $[L(\alpha):L]$ always divide $[K(\alpha):K]$?

7.6 Write down all monic cubic polynomials in $\mathbb{Z}_2[x]$, factorize them completely and construct a splitting field for each of them. Which of these fields are isomorphic?

7.7 Find a splitting field extension $K:\mathbb{Q}$ for each of the following polynomials over $\mathbb{Q}: x^4 - 5x^2 + 6$, $x^4 + 5x^2 + 6$, $x^4 - 5$. In each case determine the degree $[K:\mathbb{Q}]$ and find α such that $K = \mathbb{Q}(\alpha)$.

7.8 Find a splitting field extension $K:\mathbb{Q}$ for each of the following polynomials over $\mathbb{Q}: x^4 + 1$, $x^4 + 4$, $(x^4 + 1)(x^4 + 4)$, $(x^4 - 1)(x^4 + 4)$. In each case determine the degree $[K:\mathbb{Q}]$ and find α such that $K = \mathbb{Q}(\alpha)$.

7.9 Suppose that $L:K$ is a splitting field extension for a polynomial of degree n. Show that $[L:K]$ divides $n!$

7.10 Find a splitting field extension for $x^3 - 5$ over $\mathbb{Z}_7, \mathbb{Z}_{11}$ and $\mathbb{Z}_{13}$.

8

The algebraic closure of a field

8.1   Introduction

As we observed at the beginning of the preceding chapter, if $f \in \mathbb{Q}[x]$ we can consider f as an element of $\mathbb{C}[x]$, and then f splits over $\mathbb{C}$. We therefore have the comforting conclusion that, whenever $f \in \mathbb{Q}[x]$, we can find a splitting field extension for f which is a subfield of the fixed field $\mathbb{C}$.

In this chapter we shall show that a similar phenomenon occurs for any field K. We must make some definitions. A field L is said to be *algebraically closed* if every f in $L[x]$ splits over L. Thus the 'fundamental theorem of algebra' states that $\mathbb{C}$ is algebraically closed. An extension $L:K$ is called an *algebraic closure* of K if $L:K$ is algebraic and L is algebraically closed. Note that $\mathbb{C}:\mathbb{Q}$ is *not* an algebraic closure of $\mathbb{Q}$ since $\mathbb{C}:\mathbb{Q}$ is not algebraic (Exercise 4.7).

The next theorem gives two useful characterizations of an algebraic closure:

Theorem 8.1 *Suppose that $L:K$ is an extension. The following are equivalent:*

(i) *$L:K$ is an algebraic closure of K.*
(ii) *$L:K$ is algebraic, and every irreducible f in $K[x]$ splits over L.*
(iii) *$L:K$ is algebraic, and if $L':L$ is algebraic then $L = L'$.*

Proof. Clearly (i) implies (ii). Suppose that (ii) holds and that $L':L$ is algebraic. Then $L':K$ is also algebraic (Theorem 4.7). Suppose that $\alpha' \in L'$. Let m be the minimal polynomial of α' over K. Then m is irreducible and so, by hypothesis, m splits over L:

$$m = (x - \lambda_1)\dots(x - \lambda_n)$$

As $m(\alpha') = 0$, $\alpha' = \lambda_j$ for some j, and so $\alpha' \in L$. Thus $L = L'$ and (iii) holds.

Finally suppose that (iii) holds, and that $f \in L[x]$. By Theorem 7.3, there is a splitting field extension L' for f over L. $L':L$ is algebraic, by the corollary to Theorem 7.1 and so, by hypothesis, $L' = L$. Thus f splits over L, and so L is algebraically closed. Consequently $L:K$ is an algebraic closure of K.

Corollary *Suppose that $L:K$ is an extension and that L is algebraically closed. Let L_a be the field of elements of L which are algebraic over K. Then $L_a:K$ is an algebraic closure of K.*

In particular, if A is the field of complex numbers which are algebraic over $\mathbb{Q}$, then $A:\mathbb{Q}$ is an algebraic closure for $\mathbb{Q}$.

8.2   The existence of an algebraic closure

We now turn to the fundamental theorem concerning algebraic closures.

Theorem 8.2 *If K is a field, there exists an algebraic closure $L:K$.*

The generality of this statement suggests that we may need to use the axiom of choice, and the maximal nature of an algebraic closure revealed by Theorem 8.1 reinforces this belief. It is, however, necessary to proceed with some care. Let us begin by giving a *fallacious* argument.

Partially order the algebraic extensions $M:K$ by saying that $M_1:K \geqslant M_2:K$ if M_2 is a subfield of M_1. If $\mathscr{C}$ is a chain of extensions $M:K$, let $N = \bigcup \{M : M:K \in \mathscr{C}\}$. If $\alpha, \beta \in N$, there exists $M:K$ in $\mathscr{C}$ such that α and β are in M. Define $\alpha\beta, \alpha + \beta$ and α^{-1} (if $\alpha \neq 0$) by the operations in M. This does not depend on M, and so N is a field, and $N:K$. If $\alpha \in N$, $\alpha \in M$ for some M, and so α is algebraic over K. Thus $N:K$ is an upper bound for $\mathscr{C}$. By Zorn's lemma, there is a maximal algebraic extension, and by Theorem 8.1, this is an algebraic closure.

What is wrong with this argument? The error comes at the very beginning, when we try to compare extensions. Recall that an extension is really a triple (i, K, M), where i is a monomorphism from K into M. Thus in general we cannot compare extensions in the way that is suggested.

Nevertheless, the fallacious argument has some virtue, and it is possible, by considering fields which, as sets, are subsets of a sufficiently large fixed set, to produce a correct argument along the lines which the fallacious argument suggests. Exercises 8.1–8.3 show one way in which this can be done. We shall instead give a more 'ring-theoretic' argument, which uses the axiom of choice by appealing to Theorem 3.14.

We consider a ring of polynomials in very many variables. If f is a non-constant monic polynomial in $K[x]$ of degree n, then f has at most n roots in a splitting field extension: we introduce an indeterminate to correspond to each of these possible roots. Let U be the set of all pairs (f, j), where f is a non-constant monic polynomial in $K[x]$ and $1 \leqslant j \leqslant \text{degree } f$. For each (f, j) in U, we introduce an indeterminate $x_j(f)$, and consider the polynomial ring $K[X_U]$ of polynomials with coefficients in K and with indeterminates

$$X_U = \{x_j(f) : (f, j) \in U\}.$$

Now suppose that f is a non-constant monic polynomial in $K[x]$. We can write

$$f = x^n - a_1(f)x^{n-1} + \cdots + (-1)^n a_n(f)$$

(notice that we have not written monic polynomials in this form before: as we shall see, this can be a very useful form to use). Let $g(f)$ be the element of $K[X_U][x]$ that has $x_1(f), \ldots, x_n(f)$ as roots:

$$g(f) = \prod_{j=1}^{n} (x - x_j(f))$$

$$= x^n - s_1(f)x^{n-1} + \cdots + (-1)^n s_n(f),$$

where

$$s_j(f) = \sum_{i_1 < \cdots < i_j} x_{i_1}(f) \ldots x_{i_j}(f) \in K[X_U]$$

is the jth elementary symmetric polynomial in $x_1(f), \ldots, x_n(f)$.

The idea of the proof is to identify f and $g(f)$, and to exploit the fact that $g(f)$ splits in $K[X_U][x]$. With this in mind, we set

$$t_i(f) = s_i(f) - a_i(f)$$

for $1 \leqslant i \leqslant n$. Let I be the ideal in $K[X_U]$ generated by all the elements $t_i(f)$ as f and i vary. The main step in the proof is to show that I is a proper ideal in $K[X_U]$. For this, it is sufficient to show that $1 \notin I$; in other words to show that it is not possible to find $r_1, \ldots, r_N$ in $K[X_U]$ and elements $t_{i_1}(f_1), \ldots, t_{i_N}(f_N)$ such that

$$1 = r_1 t_{i_1}(f_1) + \cdots + r_N t_{i_N}(f_N).$$

Suppose that such an expression were to exist. By Theorem 7.3, there exists a splitting field $L:K$ for the polynomial $h = f_1 \ldots f_N$. Then each f_k splits over L; we can write

$$f_k = (x - \alpha_1(k)) \ldots (x - \alpha_{n_k}(k))$$

where $n_k = \text{degree } f_k$. Note that

$$a_j(f_k) = \sum_{i_1 < \cdots < i_j} \alpha_{i_1}(k) \ldots \alpha_{i_j}(k).$$

We now consider the evaluation map E from $K[X_U]$ to L which sends $x_i(f_k)$ to $\alpha_i(k)$ for $1 \leqslant i \leqslant n_k$ and $1 \leqslant k \leqslant N$, and which sends all the other indeterminates to 0. Then $E(s_j(f_k)) = a_j(f_k)$, and so it follows from the definition of $t_i(f)$ that

$$E(t_i(f_k)) = 0 \quad \text{for } 1 \leqslant i \leqslant n_k, \ 1 \leqslant k \leqslant N,$$

so that

$$1 = E(1) = E(r_1)E(t_{i_1}(f_1)) + \cdots + E(r_N)E(t_{i_N}(f_N)) = 0,$$

This gives the contradiction that we are looking for.

Since I is a proper ideal of $K[X_U]$, there exists a maximal proper ideal J of $K[X_U]$ which contains I, by Theorem 3.14. (This is where we use the axiom of choice.) By Theorem 3.16, $K[X_U]/J$ is a field, M say. We now let $j = qi$, where i is the natural monomorphism from K into $K[X_U]$, and q is the quotient map from $K[X_U]$ onto M:

$$K \overset{i}{\to} K[X_U] \overset{q}{\to} M.$$

Then (j, K, M) is an extension of K. Let us set $\beta_j(f) = q(x_j(f))$, for all $u = (f, j) \in U$.

Now suppose that

$$f = x^n - a_1(f)x^{n-1} + \cdots + (-1)^n a_n(f)$$

is a non-constant monic polynomial in $K[x]$. Then

$$j(f) = x^n - j(a_1(f))x^{n-1} + \cdots + (-1)^n j(a_n(f))$$

is the corresponding polynomial in $M[x]$. But

$$j(a_k(f)) = q(i(a_k(f))) = q(s_k(f)),$$

since $s_k(f) - i(a_k(f)) = t_k(f) \in I \subseteq J$. Thus

$$\begin{aligned}
j(f) &= x^n - q(s_1(f))x^{n-1} + \cdots + (-1)^n q(s_n(f)) \\
&= q(x^n - s_1(f)x^{n-1} + \cdots + (-1)^n s_n(f)) \\
&= q((x - x_1(f))(x - x_2(f)) \ldots (x - x_n(f))) \\
&= (x - \beta_1(f)) \ldots (x - \beta_n(f)),
\end{aligned}$$

and $j(f)$ splits over M. Further, each $\beta_k(f)$ is algebraic over $j(K)$ (since it is a root of $j(f)$) and the $\beta_k(f)$ generate M over K: thus $M{:}K$ is algebraic, by Corollary 2 to Theorem 4.6. Consequently (j, K, M) is an algebraic closure of K.

Exercises

8.1 (i) Suppose that U is a non-empty set, and that $P(U)$ is the set of subsets of U. Show that if $V \subseteq U$ and $f : V \to P(U)$ is a mapping, then f is not onto. (Consider $\{x : x \in V, x \notin f(x)\}$.)

(ii) Suppose that U is a non-empty set and that $V \subseteq W \subseteq U$.

Show that if $f:V \to P(U)$ is one–one then there exists a one–one map $g: W \to P(U)$ such that $g|_V = f$. (Use Zorn's lemma.)

8.2 Suppose that K is a field. Let $U = K[x] \times \mathbb{Z}^+$.

(i) Show that if (k, K, L) is an algebraic extension, then there exists a one–one mapping of L into U. (Use Zorn's lemma.)

(ii) Suppose that (k, K, L) and (l, L, M) are algebraic extensions. Show that if $f : L \to P(U)$ is one–one then there exists a one–one map $g: M \to P(U)$ such that $f = gl$.

8.3 Suppose that K is a field. Let $U = K[x] \times \mathbb{Z}^+$.

(i) If $\alpha \in K$, let $j(\alpha) = \{(x - \alpha, 1)\}$. Show that $j: K \to P(U)$ is one–one, and that $j(K)$ can be given the structure of a field in such a way that j is a field isomorphism.

(ii) Let $\mathscr{F}$ be the set of triples $(S, +, .)$ where

(a) $j(K) \subseteq S \subseteq P(U)$;

(b) $(S, +, .)$ is a field, $F(S)$ say,

(c) $(i, j(K), F(S))$ is an algebraic extension (here i is the inclusion mapping).

Define a partial order on $\mathscr{F}$ by saying that $(S_1, +_1, ._1) \leqslant (S_2, +_2, ._2)$ if $S_1 \subseteq S_2$ and $(i, F(S_1), F(S_2))$ is an extension (again, i is the inclusion mapping). Show that under this order, $\mathscr{F}$ has a maximal element (Zorn's lemma).

(iii) Use Theorem 7.2 or 7.3 to show that if $(S, +, .)$ is a maximal element of $\mathscr{F}$ then $(j, K, F(S))$ is an algebraic closure for K. (Here j is considered as a mapping of K into $F(S)$.)

8.3 The uniqueness of an algebraic closure

We now consider problems of uniqueness. First we establish an extension theorem: this uses Zorn's lemma in a very standard way.

Theorem 8.3 *Suppose that $i: K_1 \to K_2$ is a monomorphism, that $L:K_1$ is algebraic and that K_2 is algebraically closed. Then there exists a monomorphism $j:L \to K_2$ such that $j|_{K_1} = i$.*

Proof. Let S denote all pairs (M, θ), where M is a subfield of L containing K_1, and θ is a monomorphism from M into K_2 such that $\theta|_{K_1} = i$. Partially order S by setting $(M_1, \theta_1) \leqslant (M_2, \theta_2)$ if $M_1 \subseteq M_2$ and $\theta_2|_{M_1} = \theta_1$. If $\mathscr{C}$ is a chain in S, let $N = \bigcup \{M : (M, \theta) \in \mathscr{C}\}$. If $n \in N$, then $n \in M$ for some $(M, \theta) \in \mathscr{C}$. Set $\phi(n) = \theta(n)$. It is now straightforward to verify that ϕ is well defined, that $\phi : N \to K_2$ is a monomorphism and that (N, ϕ) is an upper bound for $\mathscr{C}$. Thus, by Zorn's lemma, S has a maximal element (M, θ). We must show that $M = L$.

If not, there exists $\alpha \in L \backslash M$. α is algebraic over M: let m be its minimal polynomial over M. Then $\theta(m)$ splits over K_2, since K_2 is algebraically closed. Let

$$\theta(m) = (x - \beta_1) \ldots (x - \beta_r)$$

Then $\theta(m)(\beta_1) = 0$, and so by Theorem 7.4 there exists a monomorphism $\theta_1 : M(\alpha) \to K_2$ with $\theta_1|_M = \theta$. This contradicts the maximality of (M, θ).

We are now in a position to show that an algebraic closure is essentially unique.

Theorem 8.4 *Suppose that* (i_1, K, L_1) *and* (i_2, K, L_2) *are two algebraic closures for K. Then there exists an isomorphism* $j : L_1 \to L_2$ *such that* $i_2 = ji_1$.
Proof. By Theorem 8.3 there exists a monomorphism $j : L_1 \to L_2$ such that $i_2 = ji_1$.

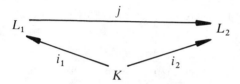

We now use Theorem 8.1. If f is irreducible over $K[x]$, $i_1(f)$ splits over L_1, and so $i_2(f)$ splits over $j(L_1)$. As $(i_2, K, j(L_1))$ is algebraic, $(i_2, K, j(L_1))$ is an algebraic closure for K. Now $L_2 : j(L_1)$ is algebraic, as (i_2, K, L_2) is, and so $L_2 = j(L_1)$, by Theorem 8.1 (iii).

In future, if K is any field, we shall denote by $\bar{K} : K$ any algebraic closure of K.

Exercises

8.4 What is the algebraic closure of $\mathbb{Q}$ (as a subfield of $\mathbb{C}$)?

8.5 Show that an algebraically closed field must be infinite.

8.6 Suppose that $K(\alpha):K$ is a simple extension and that α is transcendental over K. Show that $K(\alpha)$ is not algebraically closed.

8.7 Suppose that K is a countable field. Show how to construct an algebraic closure, by successively constructing splitting fields of the (countably many) polynomials in $K[x]$. Is your construction less fallacious than the 'fallacious proof' of Theorem 8.2?

8.8 Suppose that $L:K$ is algebraic. In what sense is it true that $\bar{L} = \bar{K}$?

8.4 Conclusions

We have now achieved what we set out to do. Some comments are in order. First, the proof of Theorem 8.2 is quite difficult. More to the point, it is quite different from the very special construction of the complex field $\mathbb{C}$. Here, the hard work is constructing the real number field $\mathbb{R}$ from the rational field $\mathbb{Q}$. $\mathbb{C}:\mathbb{R}$ is then a splitting field extension for the polynomial $x^2 + 1$, which is irreducible over $\mathbb{R}$. It is then remarkably the case that all polynomials over $\mathbb{R}$ split over $\mathbb{C}$. The complex field is a very special one!

Secondly, the proof uses the axiom of choice in an essential way. This suggests that the theorem should only be used when it is necessary to do so.

Thirdly, the existence of an algebraic closure, and the extension theorem (Theorem 8.3) provide a useful framework in which to work. If one uses this, the theory can be developed more simply in a few places. But the use of the axiom of choice seems too big a price to pay: for this reason we shall not use algebraic closures in the development of the theory.

9

Normal extensions

9.1   Basic properties

In this chapter and the next we consider two important properties that an extension may or may not have. We begin with normality.

An extension $L:K$ is said to be *normal* if it is algebraic and whenever f is an irreducible polynomial in $K[x]$ then *either* f splits over L *or* f has no roots in L. Clearly an algebraic extension $L:K$ is normal if and only if the minimal polynomial over K of each element of L splits over L.

The word 'normal' is one of the most overworked words in mathematical terminology (normal subgroups, normal topological spaces, . . .). We shall see in due course that this is a good use of the word.

In order to characterize normality, we need to extend the definition of a splitting field. Suppose that K is a field, and that S is a subset of $K[x]$. We say that an extension L of K is a *splitting field extension for S* if each f in S splits over L, and if $L \supseteq L' \supseteq K$ and each f in S splits over L', then $L' = L$.

If S is a finite set $\{f_1, \ldots, f_n\}$ then $L:K$ is a splitting field extension for S if and only if it is a splitting field extension for $g = f_1 \ldots f_n$; thus the new definition is only of interest if S is infinite.

Theorem 9.1 *An extension $L:K$ is normal if and only if it is a splitting field extension for some $S \subseteq K[x]$.*

Proof. Suppose first that $L:K$ is normal. $L:K$ is algebraic: let $S = \{m_\alpha : \alpha \in L\}$ be the set of minimal polynomials over K of elements of L. By hypothesis, each f in S splits over L, and clearly S splits over no proper subfield of L.

Conversely suppose that $L:K$ is a splitting field extension for S. Let A denote the set of roots in L of polynomials in S. Then clearly $L = K(A)$, and so $L:K$ is algebraic, by Corollary 2 to Theorem 4.6.

Suppose that $\beta \in L$ and that m is its minimal polynomial over K. We must show that m splits over L. First we reduce the problem to one concerning finite extensions. As $\beta \in K(A)$, there exist $\alpha_1, \ldots, \alpha_n$ in A such that $\beta \in K(\alpha_1, \ldots, \alpha_n)$. There exist $f_1, \ldots, f_n$ in S such that α_i is a root of f_i, for $1 \leqslant i \leqslant n$. Each f_i splits over L. Let R be the set of roots of $g = f_1 \ldots f_n$. Then $K(R):K$ is a splitting field extension for g and $\beta \in K(R)$. We now consider m as an element of $K(R)[x]$ and construct a splitting field extension $H:K(R)$ for m. Let γ be another root of m in H. We must show that in fact $\gamma \in K(R)$.

We have the following diagram, where upward pointing arrows denote inclusions:

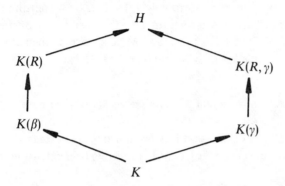

m is the minimal polynomial of both β and γ over K, so that $[K(\beta):K] = [K(\gamma):K] = \text{degree } m$. Also, by the corollary to Theorem 7.4, there is an isomorphism τ of $K(\beta)$ onto $K(\gamma)$ which sends β to γ and which fixes K. As τ fixes K, $\tau(g) = g$.

Now $K(R):K(\beta)$ is a splitting field extension for g over $K(\beta)$, and $K(R, \gamma):K(\gamma)$ is a splitting field extension for $\tau(g) = g$ over $K(\gamma)$, so that by Corollary 1 to Theorem 7.5 there is an isomorphism σ of $K(R)$ onto $K(R, \gamma)$ such that $\sigma|_{K(\beta)} = \tau$. This means that $[K(R):K(\beta)] = [K(R, \gamma):K(\gamma)]$, and so by the tower law

$$[K(R):K] = [K(R):K(\beta)][K(\beta):K]$$
$$= [K(R, \gamma):K(\gamma)][K(\gamma):K]$$
$$= [K(R, \gamma):K].$$

But $K(R) \subseteq K(R, \gamma)$, and so we must have that $K(R) = K(R, \gamma)$. Consequently, $\gamma \in K(R)$.

The case of finite extensions is particularly important:

Corollary 1 *A finite extension $L:K$ is normal if and only if $L:K$ is a splitting field extension for some $g \in K[x]$.*

For if $L:K$ is normal and finite, and $\alpha_1, \ldots, \alpha_n$ is a basis for L over K, then $L:K$ is a splitting field extension for $g = m_{\alpha_1} m_{\alpha_2} \ldots m_{\alpha_n}$.

Let $L = K(\alpha_1, \ldots, \alpha_n)$, let m_i be the minimal polynomial of α_i over K, and let $g = m_1 \ldots m_n$. Then $L:K$ is normal if and only if $L:K$ is a splitting field extension for g.

Suppose that $L:K$ is algebraic. An extension $F:L$ is a *normal closure* for $L:K$ if $F:K$ is normal, and if $F:M:L$ is a tower and $M:K$ is normal, then $M = F$.

Corollary 2 *If $L:K$ is finite, it has a finite normal closure $F:L$.*

With the same notation as in Corollary 1, let $F:L$ be a splitting field extension for g over L. Then $F:K$ is a splitting field extension for g over K so that $F:K$ is normal. If $F:M:L$ is a tower and $M:K$ is normal, then each m_α splits over M, and so g splits over M; therefore $M = F$.

Corollary 3 *If $L:K$ is normal and M is an intermediate field then $L:M$ is normal.*

For there exists $S \subseteq K[x]$ such that $L:K$ is a splitting field extension for S. If we consider S as a subset of $M[x]$, $L:M$ is a splitting field extension for S.

Exercises

9.1 Show that every algebraic extension has a normal closure.

9.2 Suppose that $L:K$ is algebraic. Show that there is a greatest intermediate field M for which $M:K$ is normal.

9.3 Suppose that $L:K$ and that M_1 and M_2 are intermediate fields. Show that if $M_1:K$ and $M_2:K$ are normal then so are $K(M_1, M_2):K$ and $M_1 \cap M_2:K$.

9.2   Monomorphisms and automorphisms

We have just seen that if $L:K$ is normal and M is an intermediate field then $L:M$ is normal. On the other hand there is no reason why $M:K$ should be normal. For example if ω is a complex cube root of 1 then $\mathbb{Q}(2^{1/3}, \omega):\mathbb{Q}$ is normal, since it is the splitting field for $f = x^3 - 2$, while $\mathbb{Q}(2^{1/3}):\mathbb{Q}$ is not, since f is irreducible and has one root in $\mathbb{Q}(2^{1/3})$ but does not split over $\mathbb{Q}(2^{1/3})$.

It is important to be able to recognize when $M:K$ is normal. In the next theorem we give necessary and sufficient conditions for this, for finite extensions: in fact the finiteness condition is not necessary (Exercises 9.6 and 9.7).

Theorem 9.2 *Suppose that L:K is a finite normal extension and that M is an intermediate field. The following are equivalent:*

(i) *M:K is normal;*

(ii) *if σ is an automorphism of L which fixes K then σ(M) ⊆ M;*

(iii) *if σ is an automorphism of L which fixes K then σ(M) = M.*

Proof. Suppose first that $M:K$ is normal, and that $σ$ is an automorphism of L which fixes K. Suppose that $α \in M$ and let m be the minimal polynomial for $α$ over K. Then $m(σ(α)) = σ(m(α)) = 0$, so that $σ(α)$ is a root of m. As m splits over M, $σ(α) \in M$, and so $σ(M) ⊆ M$. Thus (i) implies (ii).

Since $[σ(M):K] = [M:K]$ it is clear that (ii) implies (iii).

Suppose now that (iii) holds. As $L:K$ is normal, $L:K$ is the splitting field extension for some $g \in K[x]$, by Corollary 1 to Theorem 9.1. Suppose that $α \in M$. Let m be the minimal polynomial for $α$ over K. As $L:K$ is normal, m splits over L. We must show that m splits over M: that is, that all the roots of m are in M. Let $β$ be any root of m in L. By Theorem 7.4, there exists a monomorphism j from $K(α)$ to $K(β)$, fixing K, such that $j(α) = β$. Since j fixes $K, j(g) = g$. Now $L:K(α)$ and $L:K(β)$ are splitting field extensions for $j(g) = g$. By Corollary 1 to Theorem 7.5, there is an isomorphism $σ:L→L$ which extends j. As $σ$ fixes $K, σ(M) = M$. In particular, this means that $β = σ(α) \in M$.

Exercises

9.4 Suppose that $N:L$ and $N':L$ are two normal closures of $L:K$. Show that there is an isomorphism j of N onto N' such that $j(l) = l$ for $l \in L$.

9.5 Suppose that $L:K$ is a finite normal extension and that f is an irreducible polynomial in $K[x]$. Suppose that g and h are irreducible monic factors of f in $L[x]$. Show that there is an automorphism $σ$ of L which fixes K such that $σ(g) = h$.

9.6 Suppose that $L:K$ is algebraic. Show that the following are equivalent:

(i) $L:K$ is normal;

(ii) if j is any monomorphism from L to $\bar{L}$ which fixes K then $j(L) ⊆ L$;

(iii) if j is any monomorphism from L to $\bar{L}$ which fixes K then $j(L) = L$.

9.7 Show that the condition that $L:K$ is finite can be dropped from Theorem 9.2. (Use the previous exercise and Theorem 8.3.)

10

Separability

10.1 Basic ideas

The second important property that an extension may or may not have is separability. We have seen that normality is a rather special property. Separability is different: we shall have to work quite hard to produce an extension which is not separable. Lack of separability leads to technical difficulties: when the time comes, we shall avoid these by making appropriate assumptions.

Separability involves several definitions. Suppose first that f is an irreducible polynomial of degree n in $K[x]$ and that $L:K$ is a splitting field extension for f. We say that f is *separable* (over K) if f has n distinct roots in L. Suppose next that f is an arbitrary polynomial in $K[x]$. We say that f is *separable* (over K) if each of its irreducible factors is separable.

Suppose that $L:K$ is an extension and that $\alpha \in L$. We say that α is *separable* (over K) if it is algebraic over K and its minimal polynomial over K is separable, and say that $L:K$ is *separable* if each α in L is separable over K.

Theorem 10.1 *Suppose that $L:K$ is separable and that M is an intermediate field. Then $L:M$ and $M:K$ are separable.*

Proof. It is obvious that $M:K$ is separable.

Suppose that $\alpha \in L$. Let m_1 be its minimal polynomial over M, m_2 its minimal polynomial over K. Let $N:M$ be a splitting field extension for m_2, considered as an element of $M[x]$. Since m_2 is separable over K, we can write

$$m_2 = (x - \alpha_1) \ldots (x - \alpha_r)$$

where $\alpha_1, \ldots, \alpha_r$ are distinct elements of N. But $m_1 | m_2$ in $M[x]$, and so in $N[x]$

$$m_1 = (x - \alpha_{i_1}) \ldots (x - \alpha_{i_s})$$

for some $1 \leqslant i_1 < \cdots < i_s \leqslant r$. Thus m_1 is separable.

10.2 Monomorphisms and automorphisms

We have already seen that counting dimension leads to some remarkably strong results. We shall find that counting monomorphisms and automorphisms is equally useful. With this in mind, the results in this section suggest why separability is important.

First we consider simple extensions.

Theorem 10.2 *Suppose that $K(\alpha):K$ is a simple algebraic extension of degree d. Suppose that $j:K \to L$ is a monomorphism. If α is separable over K and if $j(m_\alpha)$ splits over L then there are exactly d monomorphisms from $K(\alpha)$ to L extending j; otherwise there are fewer than d such monomorphisms.*

Proof. By Corollary 2 to Theorem 7.4, there are r such extensions, where r is the number of distinct roots of $j(m_\alpha)$ in L. Now $d = \text{degree } m_\alpha = \text{degree } j(m_\alpha)$ (Theorem 4.4), so that $r \leqslant d$, and $r = d$ if and only if $j(m_\alpha)$ splits into d distinct linear factors: that is, if and only if $j(m_\alpha)$ is separable over $j(K)$ and $j(m_\alpha)$ splits over L. Clearly α is separable over K if and only if $j(m_\alpha)$ is separable over $j(K)$, and so the result is proved.

We now consider the general case.

Theorem 10.3 *Suppose that $K':K$ is a finite extension of degree d, and that $j:K \to L$ is a monomorphism. If $K':K$ is separable and $j(m_\alpha)$ splits over L for each α in K' then there are exactly d monomorphisms from K' to L extending j; otherwise, there are fewer than d such monomorphisms.*

Proof. We prove this by induction on d. It is trivially true when $d = 1$. Suppose that it is true for all extensions of degree less than d, and that $[K':K] = d$.

Suppose first that the conditions are satisfied. Let $\alpha \in K' \backslash K$. By Theorem 10.2 there are exactly $[K(\alpha):K]$ monomorphisms from $K(\alpha)$ to L extending j. Let k be one of these. We apply the inductive hypothesis to $K':K(\alpha)$. First, $[K':K(\alpha)] < d$. Secondly $K':K(\alpha)$ is separable, by Theorem 10.1. If $\beta \in K'$, let m_β be the minimal polynomial for β over K and let n_β be the minimal polynomial for β over $K(\alpha)$. Then n_β divides m_β in $K(\alpha)[x]$ and so $k(n_\beta)$ divides $k(m_\beta)$ in $L[x]$. But $k(m_\beta)$ splits over $L[x]$, and so $k(n_\beta)$ splits over $L[x]$. Thus the conditions are satisfied, and so k can be extended in $[K':K(\alpha)]$ ways. It therefore follows from the tower law that j can be extended in d ways.

Suppose next that the conditions are not satisfied. Then there exists α in K' such that $j(m_\alpha)$ has fewer than $[K(\alpha):K]$ distinct roots in L, and so j can be extended in fewer than $[K(\alpha):K]$ ways to a monomorphism from $K(\alpha)$ to L, by Corollary 2 to Theorem 7.4. Each of these extensions can be extended to a monomorphism from K' to L in at most $[K':K(\alpha)]$ ways, by the inductive hypothesis, and so there are fewer than d extensions.

Corollary 1 *Suppose that* $L:K$ *is finite and that* $L=K(\alpha_1,\ldots,\alpha_r)$. *If* α_i *is separable over* $K(\alpha_1,\ldots,\alpha_{i-1})$ *for* $1\leqslant i\leqslant r$, *then* $L:K$ *is separable.*

Proof. Let $F:L$ be a normal closure for $L:K$. Let $K_0=K$, and let $K_j=K(\alpha_1,\ldots,\alpha_j)=K_{j-1}(\alpha_j)$ for $1\leqslant j\leqslant r$. We assert that there are $[K_j:K]$ monomorphisms from K_j into F which fix K. The result is trivially true for $j=0$. Assume that it is true for $j-1$, and that i is a monomorphism from K_{j-1} to F which fixes K. Let n_j be the minimal polynomial for α_j over K_{j-1}, and let m_j be the minimal polynomial for α_j over K. Then $n_j|m_j$ in $K_{j-1}[x]$, and so $i(n_j)|i(m_j)$ in $i(K_{j-1})[x]$. But $i(m_j)=m_j$, and m_j splits in $F[x]$, so that $i(n_j)$ splits in $F[x]$. As α_j is separable over K_{j-1}, i can be extended in $[K_j:K_{j-1}]$ ways to a monomorphism from K_j to F, by Theorem 10.2. The assertion therefore follows inductively, using the tower law. But it now follows from Theorem 10.3 that $K_j:K$ is separable, and so, in particular, $L:K$ is separable.

Corollary 2 *Suppose that* $L:K$ *is finite and that* $L=K(\alpha_1,\ldots,\alpha_r)$. *If each* α_i *is separable over* K *then* $L:K$ *is separable.*

This follows from Corollary 1 and Theorem 10.1.

Corollary 3 *Suppose that* $f \in K[x]$ *is separable over* K *and that* $L:K$ *is a splitting field extension for* f. *Then* $L:K$ *is separable.*

Apply Corollary 2 to the roots of f in L.

Corollary 4 *Suppose that* $L:K$ *is finite, and that* $L:M:K$ *is a tower. If* $L:M$ *and* $M:K$ *are separable, then so is* $L:K$.

Write $M=K(\alpha_1,\ldots,\alpha_r)$, $L=M(\alpha_{r+1},\ldots,\alpha_s)$, and use Corollary 1 and Theorem 10.1.

Exercise

10.1 Suppose that $L:K$ is finite and that $L':L$ is a normal closure for $L:K$. Show that $L:K$ is separable if and only if there are exactly $[L:K]$ monomorphisms of L into L' which fix K.

10.3   Galois extensions

An extension which is finite, normal and separable is called a *Galois extension*.

If we apply Theorem 10.3 to the identity on K, we obtain the following.

Theorem 10.4 *Suppose that* $L:K$ *is finite. If* $L:K$ *is a Galois extension, there are* $[L:K]$ *automorphisms of* L *which fix* K; *otherwise there are fewer than* $[L:K]$ *such automorphisms.*

This theorem is the real starting point for Galois theory. We shall continue with this in the next chapter; in the rest of this chapter we shall study separability further.

10.4 Differentiation

Suppose that f is a non-zero element of $K[x]$ and that $L:K$ is a splitting field extension for f. We say that f has a *repeated root* in L if there exists $\alpha \in L$ and $k > 1$ such that

$$(x - \alpha)^k | f \text{ in } K[x].$$

An irreducible polynomial in $K[x]$ is not separable if and only if it has a repeated root in a splitting field. It is therefore important to be able to recognize when a polynomial has a repeated root.

Suppose that f is a non-zero polynomial in $\mathbb{C}[x]$, and that α is a root of f. How do we tell if α is a repeated root? We differentiate: α is a repeated root if and only if $f'(\alpha) = 0$. Although you have no doubt learnt about differentiation in analysis, the differential operator has strong algebraic properties – in particular, $(fg)' = f'g + fg'$ – and we can define the derivative of a polynomial in a purely algebraic way.

Suppose that

$$f = a_0 + a_1 x + \cdots + a_n x^n \in K[x].$$

We define the derivative

$$Df = a_1 + 2a_2 x + \cdots + na_n x^{n-1}.$$

Here, as usual, $ja_j = a_j + \cdots + a_j$ (j times).

D is a mapping from $K[x]$ to $K[x]$. As

$$D(f + g) = Df + Dg, \qquad D(\alpha f) = \alpha(Df),$$

D is a K-linear mapping. Also

$$D(x^m x^n) = (m + n)x^{m+n-1} = mx^{m-1}x^n + nx^m x^{n-1} = (Dx^m)x^n + x^m(Dx^n),$$

and so, by linearity,

$$D(fg) = (Df)g + f(Dg).$$

Notice also that, if K has non-zero characteristic p, then

$$Dx^p = px^{p-1} = 0.$$

Differentiation provides a test for repeated roots, just as in the case of $\mathbb{C}[x]$.

Theorem 10.5 *Suppose that f is a non-zero element of $K[x]$ and that $L:K$ is a splitting field for f. The following are equivalent:*

(i) *f has a repeated root in L;*
(ii) *there exists α in L for which $f(\alpha) = (Df)(\alpha) = 0$;*
(iii) *there exists m in $K[x]$, with degree $m \geqslant 1$, such that $m | f$ and $m | Df$.*

Proof. Suppose that f has a repeated root α in L. Then $f = (x - \alpha)^k g$, where $k > 1$ and $g \in L[x]$. Thus

$$Df = k(x - \alpha)^{k-1}g + (x - \alpha)^k Dg,$$

and so $f(\alpha) = Df(\alpha) = 0$. Thus (i) implies (ii).

Suppose that (ii) holds. Let m be the minimal polynomial of α over K. Then $m | f$ and $m | Df$, and so (iii) holds.

Suppose that (iii) holds. We can write $f = mh$, with h in $K[x]$. As f splits over L, so does m. Let α be a root of m in L. We can write $f = (x - \alpha)q$, with q in $L[x]$. Then

$$Df = q + (x - \alpha)Dq.$$

But $(x - \alpha) | Df$ in $L[x]$, since $m | Df$, and so $(x - \alpha) | q$. Thus $(x - \alpha)^2 | f$, and f has a repeated root in L.

This theorem enables us to characterize irreducible polynomials which are not separable.

Theorem 10.6 *Suppose that $f \in K[x]$ is irreducible. Then f is not separable if and only if* char $K = p > 0$ *and f has the form*

$$f = a_0 + a_1 x^p + a_2 x^{2p} + \cdots + a_n x^{np}.$$

Proof. If f is not separable, there exists m in $K[x]$, with degree $m \geqslant 1$, such that $m | f$ and $m | Df$. As f is irreducible, f and m are associates. Thus $f | Df$; as degree $Df <$ degree f, it follows that $Df = 0$. This can only happen if char $K \neq 0$ and f has the form given in the theorem.

Conversely, if the conditions are satisfied, $Df = 0$ and we can take $f = m$ in Theorem 10.5(iii).

Corollary *If* char $K = 0$, *all polynomials in $K[x]$ are separable.*

Theorem 10.6 raises the question: if char $K = p > 0$, and f has the form

$$f = a_0 + a_1 x^p + \cdots + a_n x^{np},$$

when is f irreducible in $K[x]$? Before answering this, we introduce an idea which has many applications.

Exercises

10.2 Suppose that f is a polynomial in $K[x]$ of degree n and that either char $K = 0$ or char $K > n$. Suppose that $\alpha \in K$. Establish *Taylor's formula*:

$$f = f(\alpha) + Df(\alpha)(x - \alpha) + \frac{D^2 f(\alpha)}{2!}(x - \alpha)^2 + \cdots + \frac{D^n f(\alpha)}{n!}(x - \alpha)^n.$$

10.3 Suppose that f is a polynomial in $K[x]$ of degree n and that either char $K = 0$ or char $K > n$. Show that α is a root of exact multiplicity $r(\leqslant n)$ if and only if

$$f(\alpha) = Df(\alpha) = \cdots = D^{r-1}f(\alpha) \quad \text{and} \quad D^r f(\alpha) \neq 0.$$

10.5 The Frobenius monomorphism

Theorem 10.6 shows that if we are to find an inseparable polynomial we must consider fields of non-zero characteristic. The next result is particularly useful for dealing with these.

Theorem 10.7 Suppose that char $K = p \neq 0$. The map $\phi(\alpha) = \alpha^p$ is a monomorphism of K into itself. The set of elements which remain fixed under ϕ is exactly the prime subfield.

Proof. Of course $\phi(\alpha\beta) = \phi(\alpha)\phi(\beta)$ and $\phi(1) = 1$. As usual, if $n \in \mathbb{Z}^+$ and $\alpha \in K$ let $n\alpha = \alpha + \cdots + \alpha$ (n times). The standard inductive argument shows that the binomial theorem holds in K, and so

$$(\alpha + \beta)^p = \alpha^p + \binom{p}{1}\alpha^{p-1}\beta + \cdots + \binom{p}{p-1}\alpha\beta^{p-1} + \beta^p.$$

But, as p is prime, $p | \binom{p}{r}$ for $1 \leqslant r < p$, and so

$$\binom{p}{r}\alpha^{p-r}\beta^r = 0.$$

Thus $(\alpha + \beta)^p = \alpha^p + \beta^p$ and ϕ is a monomorphism. The set of elements fixed by ϕ is a subfield, and therefore contains the prime subfield. But α is fixed by ϕ if and only if α is a root of $x^p - x$, and so at most p elements are fixed by ϕ. Since the prime subfield has p elements, it must be the set of elements fixed by ϕ.

The mapping ϕ is called the *Frobenius monomorphism*.

Corollary If char $K = p \neq 0$ and K is algebraic over its prime subfield, then the Frobenius monomorphism ϕ is an automorphism.

This is an immediate consequence of Theorem 4.8.

Exercises

10.4 Suppose that p is a prime number. By factorizing $x^{p-1} - 1$ over $\mathbb{Z}_p$, show that $(p-1)! + 1 = 0 \pmod{p}$ (Wilson's theorem).

10.5 Suppose that p is a prime number of the form $4n + 1$. Show that there exists k such that $k^2 + 1 = 0 \pmod{p}$. Show that p is not a prime in $\mathbb{Z} + i\mathbb{Z}$ and show that there exist u and v in $\mathbb{Z}$ such that $u^2 + v^2 = p$.

10.6 Suppose that p is a prime number of the form $4n+3$. Show that p is a prime in $\mathbb{Z}+i\mathbb{Z}$.

10.6   Inseparable polynomials
Suppose that
$$f = a_0 + a_1 x^p + \cdots + a_n x^{np}$$
is in $K[x]$. We shall write $f(x) = g(x^p)$, where
$$g = a_0 + a_1 x + \cdots + a_n x^n.$$
This is a slight abuse of terminology, which does not lead to any difficulties.

Theorem 10.8 *Suppose that* char $K = p > 0$ *and that*
$$f(x) = g(x^p) = a_0 + a_1 x^p + \cdots + x^{np}$$
is monic; then f is irreducible in $K[x]$ if and only if g is irreducible in $K[x]$, and not all of the coefficients a_i are pth powers of elements of K.
Proof. If g factorizes as $g = g_1 g_2$, then f factorizes as $f(x) = g_1(x^p)g_2(x^p)$: thus if f is irreducible, so is g.

Suppose next that each a_i is a pth power of an element of K: that is, $a_i = b_i^p$, for b_i in K. Then
$$f = b_0^p + b_1^p x^p + \cdots + b_n^p x^{np}$$
$$= (b_0 + b_1 x + \cdots + b_n x^n)^p$$
and so f factorizes. Thus if f is irreducible, not all the a_i can be pth powers of elements of K.

Conversely, suppose that f factorizes. We must show that either g factorizes or that all the a_i are pth powers of elements of K. We can write f as a product of irreducible factors:
$$f = f_1^{n_1} \ldots f_r^{n_r}$$
where the f_i are monic and irreducible in $K[x]$, f_i and f_j are relatively prime, for $i \neq j$, and $n_1 + \cdots + n_r > 1$. We have to consider two cases.

First suppose that $r > 1$. Then we can write $f = h_1 h_2$, with h_1 and h_2 relatively prime (take $h_1 = f_1^{n_1}$).
There exist λ_1 and λ_2 in $K[x]$ such that
$$\lambda_1 h_1 + \lambda_2 h_2 = 1.$$
Further,
$$0 = Df = (Dh_1)h_2 + h_1(Dh_2).$$
Eliminating h_2, we find that
$$Dh_1 = \lambda_1 h_1(Dh_1) - \lambda_2 h_1(Dh_2)$$
and so $h_1 | Dh_1$. As degree $Dh_1 <$ degree h_1, we must have $Dh_1 = 0$. Similarly

$Dh_2 = 0$. Thus we can write

$$h_1(x) = c_0 + c_1 x^p + \cdots + c_s x^{sp} = j_1(x^p),$$
$$h_2(x) = d_0 + d_1 x^p + \cdots + d_t x^{tp} = j_2(x^p)$$

and g factorizes as $g = j_1 j_2$.

Secondly, suppose that $r = 1$. Then $f = f_1^n$, where f_1 is irreducible, and $n > 1$. Again there are two cases to consider. If $p|n$, we can write $f = h^p$. If

$$h = c_0 + c_1 x + \cdots + c_s x^s$$

then, arguing as in Theorem 10.7,

$$f = h^p = c_0^p + c_1^p x^p + \cdots + c_s^p x^{sp}$$

so that all the a_i are pth powers. If p does not divide n,

$$0 = Df = n(Df_1)f_1^{n-1}$$

and so $Df_1 = 0$. Thus we can write

$$f_1(x) = d_0 + d_1 x^p + \cdots + d_t x^{tp} = g_1(x^p)$$

and $g = (g_1)^n$.

Corollary *If* char $K = p \neq 0$ *and K is algebraic over its prime subfield, then all polynomials in $K[x]$ are separable.*

Proof. As the Frobenius monomorphism maps K onto K, every element of K is a pth power. If f is irreducible, f can therefore not be of the form $f(x) = g(x^p)$, and so f must be separable, by Theorem 10.6.

Bearing in mind the corollary to Theorem 10.6, this means that if we are to find an inseparable polynomial we must consider fields K of non-zero characteristic which are not algebraic over their prime subfields.

With this information, the search is rather short. Let $K = \mathbb{Z}_p(\alpha)$ be the field of rational expressions in α over $\mathbb{Z}_p$. Suppose if possible that $-\alpha = \beta^p$, for some β in K. Then we can write $\beta = f(\alpha)/g(\alpha)$, with f and g in $\mathbb{Z}_p[x]$. Thus

$$-\alpha(g(\alpha))^p = (f(\alpha))^p$$

and so, since α is transcendental,

$$-xg^p = f^p.$$

But $p|\mathrm{degree}\,(f^p)$ and p does not divide degree $(-xg^p)$. Thus $-\alpha$ is not a pth power in K, and so $x^p - \alpha$ is irreducible in $K[x]$, by Theorem 10.8. Let $L:K$ be a splitting field extension for $x^p - \alpha$, and let γ be a root of $x^p - \alpha$ in L. Then

$$(x - \gamma)^p = x^p - \gamma^p = x^p - \alpha$$

so that $x^p - \alpha$ fails to be separable in the most spectacular way.

Exercises

10.7 Suppose that char $K = p \neq 0$. Show that every polynomial in $K[x]$ is separable (K is *perfect*) if and only if the Frobenius monomorphism is an automorphism of K.

10.8 Show that a field K is perfect if and only if every finite extension of K is separable.

10.9 Suppose that char $K = p > 0$ and that f is irreducible in $K[x]$. Show that f can be written in the form $f(x) = g(x^{p^n})$, where n is a non-negative integer and g is irreducible and separable.

10.10 Suppose that char $K = p > 0$ and that $L:K$ is a *totally inseparable* algebraic extension: that is, every element of $L \backslash K$ is inseparable. Show that if $\beta \in L$ then its minimal polynomial over K is of the form $x^{p^n} - \alpha$, where $\alpha \in K$.

10.11 Suppose that char $K = p \neq 0$, that f is irreducible in $K[x]$ and that $L:K$ is a splitting field extension for f. Show that there exists a non-negative integer n such that every root of f in L has multiplicity p^n. (Hint: use Exercise 10.3.)

11

Automorphisms and fixed fields

Galois theory is largely concerned with properties of groups of automorphisms of a field. If L is a field, we denote by Aut L the set of all automorphisms of L. Aut L is a group under the usual law of composition.

Suppose that A is a subset of Aut L. We set

$$\phi(A) = \{k \in L : \sigma(k) = k \text{ for each } \sigma \text{ in } A\}.$$

It is easy to verify that $\phi(A)$ is a subfield of L, which we call the *fixed field* of A. In this way, starting from A we obtain an extension $L:\phi(A)$.

Conversely suppose that $L:K$ is an extension. We denote by $\Gamma(L:K)$ the set of those automorphisms of L which fix K:

$$\Gamma(L:K) = \{\sigma \in \text{Aut } L : \sigma(k) = k \text{ for all } k \text{ in } K\}.$$

When there is no doubt what the larger field L is, we shall write $\gamma(K)$ for $\Gamma(L:K)$. It is again easy to verify that $\Gamma(L:K)$ is a subgroup of Aut L; we call $\Gamma(L:K)$ the *Galois group* of the extension $L:K$. In this case, then, starting from an extension we obtain a set of automorphisms.

In this chapter we shall study this reciprocal relationship in detail.

11.1   Fixed fields and Galois groups

The operations $A \to \phi(A)$ and $L:K \to \gamma(K)$ establish a *polarity* between sets of automorphisms of L and extensions $L:K$. The next theorem is a standard result for such polarities.

Theorem 11.1 *Suppose that $L:K$ is an extension, and that A is an subset of* Aut L.

   (i) $\gamma\phi(A) \supseteq A$;

  (ii) $\phi\gamma(K) \supseteq K$;

 (iii) $\phi\gamma\phi(A) = \phi(A)$;

 (iv) $\gamma\phi\gamma(K) = \gamma(K)$.

Proof. If $\sigma \in A$, $\sigma(k) = k$ for each k in $\phi(A)$, so that $\sigma \in \gamma\phi(A)$: this establishes (i). If $k \in K$, $\sigma(k) = k$ for each σ in $\gamma(K)$, so that $k \in \phi\gamma(K)$: this establishes (ii).

If $A_1 \subseteq A_2$, then clearly $\phi(A_1) \supseteq \phi(A_2)$. Thus it follows from (i) that

$$\phi\gamma\phi(A) \subseteq \phi(A):$$

but applying (ii), with $\phi(A)$ in place of K,

$$\phi\gamma\phi(A) \supseteq \phi(A).$$

This establishes (iii).

Similarly if $K_1 \subseteq K_2$, $\gamma(K_1) \supseteq \gamma(K_2)$. Applying this to (ii):

$$\gamma\phi\gamma(K) \subseteq \gamma(K)$$

but applying (i), with $\gamma(K)$ in place of A,

$$\gamma\phi\gamma(K) \supseteq \gamma(K).$$

This establishes (iv).

Corollary *If A is a subset of* Aut L, *and $\langle A \rangle$ is the subgroup of* Aut L *generated by A, then $\phi(A) = \phi(\langle A \rangle)$.*

For $A \subseteq \langle A \rangle \subseteq \gamma\phi(A)$, by (i), and so

$$\phi(A) \supseteq \phi(\langle A \rangle) \supseteq \phi\gamma\phi(A) = \phi(A), \text{ by (iii).}$$

Because of this, we shall usually restrict attention to *subgroups* of Aut L.

Suppose now that G is a subgroup of Aut L. If $\lambda \in L$, we define the *trajectory* of λ, $T(\lambda)$, to be the element of L^G defined by $T(\lambda)(\sigma) = \sigma(\lambda)$. L^G is a vector space over L; we can also consider it as a vector space over any subfield of L, and in particular as a vector space over $\phi(G)$.

The next theorem is particularly important: it takes a rather curious form, as it is concerned with linear independence over two different fields.

Theorem 11.2 *Suppose that G is a subgroup of* Aut L, *that K is the fixed field of G and that B is a subset of L. Then the following are equivalent:*

(i) *B is linearly independent over K;*

(ii) *$\{T(\beta) : \beta \in B\}$ is linearly independent over K;*

(iii) *$\{T(\beta) : \beta \in B\}$ is linearly independent over L.*

Proof. Clearly (iii) implies (ii). Suppose that B is not linearly independent over K: there exist distinct $\beta_1, \ldots, \beta_n$ in B, and $k_1, \ldots, k_n$ in K, not all zero, such that

$$k_1\beta_1 + \cdots + k_n\beta_n = 0.$$

Then if $\sigma \in G$,

$$k_1\sigma(\beta_1) + \cdots + k_n\sigma(\beta_n) = \sigma(k_1\beta_1 + \cdots + k_n\beta_n) = 0,$$

so that $k_1 T(\beta_1) + \cdots + k_n T(\beta_n) = 0$, and the set $\{T(\beta) : \beta \in B\}$ is not linearly independent over K in L^G. Thus (ii) implies (i).

Finally, suppose that the set of trajectories $\{T(\beta): \beta \in B\}$ is not linearly independent over L in L^G. There exist $\beta_1, \ldots, \beta_r$ in B, and non-zero $\lambda_1, \ldots, \lambda_r$ in L such that

$$\lambda_1 T(\beta_1) + \cdots + \lambda_r T(\beta_r) = 0;$$

further we can find $\beta_1, \ldots, \beta_r$ and $\lambda_1, \ldots, \lambda_r$ so that r is as small as possible. In detail, this says that

$$\lambda_1 \sigma(\beta_1) + \cdots + \lambda_r \sigma(\beta_r) = 0 \text{ for each } \sigma \text{ in } G. \tag{*}$$

Now if $\tau \in G$ and $\sigma \in G$ then $\tau^{-1}\sigma \in G$, so that

$$\lambda_1 \tau^{-1}\sigma(\beta_1) + \cdots + \lambda_r \tau^{-1}\sigma(\beta_r) = 0 \text{ for each } \sigma \text{ in } G.$$

Operate on this equation by τ:

$$\tau(\lambda_1)\sigma(\beta_1) + \cdots + \tau(\lambda_r)\sigma(\beta_r) = 0 \text{ for each } \sigma \text{ in } G. \tag{**}$$

Now multiply (*) by $\tau(\lambda_r)$, (**) by λ_r, and subtract:

$$(\tau(\lambda_r)\lambda_1 - \tau(\lambda_1)\lambda_r)\sigma(\beta_1) + \cdots + (\tau(\lambda_r)\lambda_{r-1} - \tau(\lambda_{r-1})\lambda_r)\sigma(\beta_{r-1}) = 0$$

for each σ in G. Thus

$$(\tau(\lambda_r)\lambda_1 - \tau(\lambda_1)\lambda_r)T(\beta_1) + \cdots + (\tau(\lambda_r)\lambda_{r-1} - \tau(\lambda_{r-1})\lambda_r)T(\beta_{r-1}) = 0.$$

Since there are fewer than r terms in the relationship, it follows from the minimality of r that all the coefficients must be zero:

$$\tau(\lambda_r)\lambda_i = \tau(\lambda_i)\lambda_r \text{ for } 1 \leqslant i < r;$$

in other words,

$$\tau(\lambda_r^{-1}\lambda_i) = \lambda_r^{-1}\lambda_i \text{ for } 1 \leqslant i < r.$$

Now this holds for each τ in G, and so $k_i = \lambda_r^{-1}\lambda_i \in K$, for $1 \leqslant i < r$. Multiplying (*) by λ_r^{-1}, we obtain

$$k_1 \sigma(\beta_1) + \cdots + k_{r-1}\sigma(\beta_{r-1}) + \sigma(\beta_r) = 0$$

for each σ in G. But as G is a subgroup of Aut L, the identity automorphism is in G. Thus

$$k_1 \beta_1 + \cdots + k_{r-1}\beta_{r-1} + \beta_r = 0$$

and so B is not linearly independent over K. Thus (i) implies (iii).

The next theorem shows that when G is finite we can relate the order of G to the degree of $L:\phi(G)$ in a most satisfactory way.

Theorem 11.3 *Suppose that G is a finite subgroup of* Aut L. *Then* $|G| = [L:\phi(G)]$, $G = \gamma\phi(G)$ *and* $L:\phi(G)$ *is a Galois extension.*
Proof. Let $K = \phi(G)$. If B is a subset of L which is linearly independent over K then, by Theorem 11.2, $\{T(\beta): \beta \in B\}$ is a subset of L^G which is linearly independent over L. But L^G has dimension $|G|$, and so $|B| \leqslant |G|$. Thus L is finite dimensional over K, and $[L:K] \leqslant |G|$. On the other hand, by Theorem

10.4, $|\gamma\phi(G)| \leqslant [L:K]$. As $G \subseteq \gamma\phi(G)$, it follows that $[L:K]=|G|$ and that $G=\gamma\phi(G)$. Since $[L:K]=|G|$, it follows from Theorem 10.4 that $L:K$ is a Galois extension.

What happens if, instead of starting with a group of automorphisms, we start with a finite extension? Here the results are not quite so clear cut. Once again, Theorem 10.4 plays a decisive rôle.

Theorem 11.4 *Suppose that $L:K$ is finite. If $L:K$ is a Galois extension, then $|\gamma(K)|=[L:K]$, and $K=\phi\gamma(K)$. Otherwise, $|\gamma(K)| < [L:K]$ and K is a proper subfield of $\phi\gamma(K)$.*
Proof. The relationship between $|\gamma(K)|$ and $[L:K]$ is given by Theorem 10.4. By Theorem 11.3, $|\gamma(K)|=[L:\phi\gamma(K)]$. Thus, if $L:K$ is normal and separable,

$$[L:K]=[L:\phi\gamma(K)];$$

as $K \subseteq \phi\gamma(K)$, $K=\phi\gamma(K)$. Otherwise

$$[L:K] > [L:\phi\gamma(K)]$$

so that K is a proper subfield of $\phi\gamma(K)$.

Exercises

11.1 Suppose that $L:K$ is a Galois extension with Galois group G, and that $\alpha \in L$. Show that $L = K(\alpha)$ if and only if the images of α under G are all distinct.

11.2 Suppose that $L:K$ is an extension. If $\sigma \in \Gamma(L:K)$, $\sigma \in \mathrm{End}_K(L)$, the K-linear space of K-linear mappings of L into itself. Show that $\Gamma(L:K)$ is a linearly independent subset of $\mathrm{End}_K(L)$.

11.3 Suppose that $L:K$ is a Galois extension with Galois group $G = \{\sigma_1,\dots,\sigma_n\}$. Show that $(\beta_1,\dots,\beta_n)$ is a basis for L over K if and only if $\det(\sigma_i(\beta_j)) \neq 0$.

11.4 Suppose that char $K=0$ and that $L:K$ is a finite extension; let $\beta_1,\dots,\beta_n$ be a basis for L over K. Suppose that H is a subgroup of $\Gamma(L:K)$; let $\gamma_j = \sum_{\sigma \in H} \sigma\beta_j$, for $1 \leqslant j \leqslant n$. Show that $K(\gamma_1,\dots,\gamma_n)$ is the fixed field for H.

11.2   The Galois group of a polynomial

The main purpose of the theory of field extensions is to deal with polynomials and their splitting fields.

Suppose that $f \in K[x]$ and that $L:K$ is a splitting field extension for f over K. Then we call $\Gamma(L:K)$ the *Galois group* of f; we denote it by $\Gamma_K(f)$ (or

$\Gamma(f)$, when it is clear what K is). By Corollary 1 to Theorem 7.5, $\Gamma_K(f)$ depends on f and K, but not on any particular choice of splitting field.

Let us interpret Theorem 11.4 in this setting.

Theorem 11.5 *Suppose that $f \in K[x]$ and that $L:K$ is a splitting field extension for f. If f is separable then $|\Gamma(f)| = [L:K]$ and $K = \phi(\Gamma(f))$; otherwise $|\Gamma(f)| < [L:K]$ and K is a proper subfield of $\phi(\Gamma(f))$.*

An element σ of $\Gamma(f)$ is an automorphism of L; it is the action of σ on the roots of f that is all important. The next result shows that we lose no information if we concentrate on this action.

Theorem 11.6 *Suppose that $f \in K[x]$ and that $L:K$ is a splitting field extension for f over K. Let R denote the set of roots of f in L. Each σ in $\Gamma(f)$ defines a permutation of R, so that we have a mapping from $\Gamma(f)$ into the group Σ_R of permutations of R. This mapping is a group homomorphism, and is one–one.*

Proof. If $\sigma \in \Gamma(f)$, then $\sigma(f) = f$, since f has its coefficients in K. Thus, if $\alpha \in R$,

$$f(\sigma(\alpha)) = \sigma(f)(\sigma(\alpha)) = \sigma(f(\alpha)) = \sigma(0) = 0.$$

Thus σ maps R into R. Since σ is one–one and R is finite, $\sigma|_R$ is a permutation. By definition,

$$(\sigma_1 \sigma_2)(\alpha) = \sigma_1(\sigma_2(\alpha))$$

so that the mapping: $\sigma \to \sigma|_R$ is a group homomorphism. Finally, if $\sigma(\alpha) = \tau(\alpha)$ for each α in R, then $\sigma^{-1}\tau$ fixes $K(R) = L$, so that $\sigma = \tau$.

Notice that Corollary 2 to Theorem 7.5 states that, if f is irreducible, then $\Gamma(f)$ acts *transitively* on the roots of f: if α and β are two roots of f in a splitting field, there exists σ in $\Gamma(f)$ with $\sigma(\alpha) = \beta$.

Conversely, suppose that f is a monic polynomial of degree n in $K[x]$ which has n distinct roots in a splitting field L, and that $\Gamma(f)$ acts transitively on the roots of f. Let α be a root of f, and let m be the minimal polynomial of α. Then if β is any root of f there exists σ in $\Gamma(f)$ such that $\sigma(\alpha) = \beta$. Thus

$$m(\beta) = m(\sigma(\alpha)) = \sigma(m)(\sigma(\alpha)) = \sigma(m(\alpha)) = 0,$$

and so m has at least n roots. Since m divides f, $m = f$ and it follows that f is irreducible.

Exercises

11.5 Describe the transitive subgroups of Σ_3, Σ_4 and Σ_5.

11.6 Find the Galois group of $x^4 - 2$ over (a) the rational field $\mathbb{Q}$, (b) the field $\mathbb{Z}_3$ and (c) the field $\mathbb{Z}_7$.

11.7 Find the Galois group of $x^4 + 2$ over (a) the rational field $\mathbb{Q}$, (b) the field $\mathbb{Z}_3$ and (c) the field $\mathbb{Z}_5$.

11.3   An example

Let us give an example. First we need some results about permutation groups. Suppose that X is a set and that G is a subgroup of the permutation group Σ_X. We define a relation on X by setting $x \sim y$ if either $x = y$ or the transposition (x, y) is an element of G. This relation is clearly reflexive and symmetric. As

$$(x, y)(y, z)(x, y) = (x, z),$$

it is also transitive: thus it is an equivalence relation on X.

Suppose now that X is finite and that G acts transitively on X. Suppose that E_x and E_y are distinct equivalence classes. As G acts transitively, there exists σ in G such that $\sigma(x) = y$. Now, if $x' \in E_x$,

$$\sigma(x, x')\sigma^{-1} = (\sigma(x), \sigma(x')) = (y, \sigma(x')) \in G$$

and so $\sigma(x') \in E_y$. Thus $\sigma(E_x) \subseteq E_y$, and so $|E_x| \leqslant |E_y|$. Similarly $|E_y| \leqslant |E_x|$, and so any two equivalence classes have the same number of elements.

In particular if X has a prime number of elements, if G acts transitively on X and if G contains at least one transposition, then there can only be one equivalence class, namely the whole of X, and so G contains *all* transpositions. As the transpositions generate Σ_X, it follows that $G = \Sigma_X$.

We now come to our example.

Theorem 11.7 *Suppose that $f \in \mathbb{Q}[x]$ is irreducible and has prime degree p. If f has exactly $p - 2$ real roots and 2 complex roots in $\mathbb{C}$ then the Galois group $\Gamma(f)$ of f over $\mathbb{Q}$ is Σ_p.*

Proof. Let $L:\mathbb{Q}$ be a splitting field extension for f, with $L \subseteq \mathbb{C}$. As f is irreducible, $\Gamma(f)$ acts transitively on the roots of f. Also the automorphism $z \to \bar{z}$ of $\mathbb{C}$ fixes the real roots of f and interchanges the complex ones (if α is a root of f, $f(\bar{\alpha}) = \bar{f}(\bar{\alpha}) = \overline{f(\alpha)} = 0$, so that $\bar{\alpha}$ is a root of f): since L is generated by the roots of f over $\mathbb{Q}$, $\bar{L} = L$, and $\Gamma(f)$ contains a transposition. The result therefore follows from the remarks preceding the statement of the theorem.

As a concrete example, let us consider

$$f = x^5 - 4x + 2.$$

f is irreducible over $\mathbb{Q}$, by Eisenstein's criterion. The function $t \to f(t)$ on $\mathbb{R}$ is continuous and differentiable, and so, by Rolle's theorem, between any two real zeros of f there is a zero of f'. But

$$f' = 5x^4 - 4$$

has only two real zeros, so that f has at most three real zeros. As

$$f(-2)=-22, \quad f(0)=2, \quad f(1)=-1, \quad f(2)=26$$

f has at least three real roots, by the intermediate value theorem. Thus f has three real roots and two complex roots; by the theorem, $\Gamma(f)=\Sigma_5$. Notice how useful elementary analysis can be!

Exercise

11.8 Sketch the graph of the polynomial

$$f_r=(x^2+4)x(x^2-4)(x^2-16)\ldots(x^2-4r^2).$$

Show that if k is an odd integer then $|f_r(k)| \geqslant 5$. Show that f_r-2 is irreducible, and determine its Galois group over $\mathbb{Q}$ when $2r+3$ is a prime.

11.4   The fundamental theorem of Galois theory

The fundamental theorem of Galois theory describes in some detail the polarity that was introduced at the beginning of the chapter.

Theorem 11.8 *Suppose that $L:K$ is finite. Let $G=\Gamma(L:K)$, and let $K_0=\phi(G)$. If $L:M:K_0$, let $\gamma(M)=\Gamma(L:M)$.*

(i) *The map ϕ is a one–one map from the set of subgroups of G onto the set of fields M intermediate between L and K_0. γ is the inverse map.*

(ii) *A subgroup H of G is normal if and only if $\phi(H):K_0$ is a normal extension.*

(iii) *Suppose that $H \lhd G$. If $\sigma \in G$, $\sigma|_{\phi(H)} \in \Gamma(\phi(H):K_0)$. The map $\sigma \to \sigma|_{\phi(H)}$ is a homomorphism of G onto $\Gamma(\phi(H):K_0))$, with kernel H. Thus*

$$\Gamma(\phi(H):K_0) \cong G/H.$$

Proof. (i) If H is a subgroup of G, H is finite, and so $\gamma\phi(H)=H$ (Theorem 11.3). Thus ϕ is one–one. $L:K_0$ is a Galois extension of K_0 (Theorem 11.3); thus if $L:M:K_0$, $L:M$ is normal (Corollary 3 of Theorem 9.1) and separable (Theorem 10.1). By Theorem 11.4, $\phi\gamma(M)=M$. Thus ϕ is onto, and γ is the inverse mapping.

(ii) Suppose that $L:M:K_0$, and that $\sigma \in G$. Then $L:\sigma(M):K_0$ (since $\sigma(L)=L$, $\sigma(K_0)=K_0$). We shall show that $\gamma(\sigma(M))=\sigma(\gamma(M))\sigma^{-1}$. For $\tau \in \gamma(\sigma(M))$ if and only if $\tau\sigma(m)=\sigma(m)$ for each m in M, and this happens if and only if $\sigma^{-1}\tau\sigma(m)=m$ for each m in M. Thus $\tau \in \gamma(\sigma(M))$ if and only if $\sigma^{-1}\tau\sigma \in \gamma(M)$: that is, if and only if $\tau \in \sigma(\gamma(M))\sigma^{-1}$.

Suppose that $H \lhd G$. Then, if $\sigma \in G$, using (i)

$$H = \sigma H \sigma^{-1} = \sigma(\gamma\phi(H))\sigma^{-1} = \gamma(\sigma(\phi(H))).$$

Applying ϕ, $\phi(H) = \phi\gamma(\sigma(\phi(H))) = \sigma(\phi(H))$ for each σ in G. By Theorem 9.2, $\phi(H):K_0$ is normal.

Conversely if $\phi(H):K_0$ is normal, then

$$\phi(H) = \sigma\phi(H) \text{ for each } \sigma \text{ in } G,$$

by Theorem 9.2. Thus

$$H = \gamma\phi(H) = \gamma(\sigma(\phi(H))) = \sigma(\gamma\phi(H))\sigma^{-1} = \sigma H \sigma^{-1}$$

for each σ in G, and so $H \lhd G$.

(iii) Suppose now that $H \lhd G$, so that $\phi(H):K_0$ is normal. If $\sigma \in G$, $\sigma(\phi(H)) = \phi(H)$, by Theorem 9.2. Thus $\sigma|_{\phi(H)}$ is an automorphism of $\phi(H)$ fixing K_0: that is, an element of $\Gamma(\phi(H):K_0)$. Since the group multiplication is the composition of mappings, the mapping $\sigma \to \sigma|_{\phi(H)}$ is a homomorphism of G into $\Gamma(\phi(H):K_0)$. σ is the kernel of this homomorphism if and only if $\sigma|_{\phi(H)}$ is the identity: that is, if and only if σ fixes $\phi(H)$. Thus the kernel is $\gamma\phi(H) = H$. Finally if $\rho \in \Gamma(\phi(H):K_0)$, there exists a monomorphism $\sigma: L \to L$ which extends ρ (Theorem 10.3). As $[L:\phi(H)] = [\sigma(L):\sigma\phi(H)] = [\sigma(L):\phi(H)]$, $\sigma(L) = L$ and σ is an automorphism. As ρ fixes K_0, σ is in G; as $\sigma|_{\phi(H)} = \rho$, the homomorphism maps G onto $\Gamma(\phi(H):K_0)$.

This completes the proof. A few remarks are in order. First, the proof is, to a large extent, a case of putting together results which have been established earlier. It is worth pausing, and tracing these results back, to their sources: this frequently turns out to be Theorem 7.4. Secondly, the theorem relates subgroups of G to intermediate fields: order is reversed and, the smaller the subgroup, the larger is the intermediate field. *All* the subgroups of G relate to *all* the intermediate fields M of $L:K_0$. Remember that this occurs in terms of $\Gamma(L:M)$, and *not* $\Gamma(M:K_0)$. *Normal* subgroups relate to intermediate fields M for which $M:K_0$ is *normal*; this justifies the terminology. If $M:K_0$ is normal, we can calculate $\Gamma(M:K_0)$ in terms of $\Gamma(L:K)$, but as a *quotient*, not as a *subgroup*. Finally, we do not need normality or separability. But if $L:K$ is a Galois extension, then $K = K_0$, and the result is correspondingly neater.

Exercises

11.9 Given a finite group G show that there exists a Galois extension $L:K$ such that $\Gamma(L:K) \cong G$.

11.10 Suppose that K_1 and K_2 are subfields of a field L such that $L:K_1$ and $L:K_2$ are both Galois extensions, with Galois groups G_1 and

G_2 respectively. Show that $L:K_1 \cap K_2$ is a Galois extension if and only if G, the group generated by G_1 and G_2, is finite, and that if this is so then $G = \Gamma(L:K_1 \cap K_2)$.

11.11 To answer this question, you need the following fact from group theory:

If G is a finite group and $|G| = p^r q$ where p is a prime which does not divide q then G has subgroups of order p^s for $1 \leqslant s \leqslant r$. (A subgroup of order p^r is called a *Sylow p-subgroup*.)

Suppose that $L:K$ is an extension with $[L:K] = 2$, that every element of L has a square root in L, that every polynomial of odd degree in $K[x]$ has a root in K and that char $K \neq 2$. Let f be an irreducible polynomial in $K[x]$, let $M:L$ be a splitting field extension for f over L, let $G = \Gamma(M:K)$ and let $H = \Gamma(M:L)$.

   (i) By considering the fixed field of a Sylow 2-subgroup of G, show that $|G| = 2^n$.
   (ii) By considering a subgroup of index 2 in H, show that if $n > 1$ then there is an irreducible quadratic in $L[x]$.
   (iii) Show that L is algebraically closed.
   (iv) Show that the complex numbers are algebraically closed.

11.12 By considering the splitting field of all polynomials of odd degree over $\mathbb{Z}_2$, show that the condition char $K \neq 2$ cannot be dropped from Exercise 11.11.

11.5 The theorem on natural irrationalities

Suppose that $f \in K[x]$ has Galois group $\Gamma_K(f)$, and that $L:K$ is an extension. Then we can consider f as an element of $L[x]$, and can consider the Galois group $\Gamma_L(f)$. In each case, we can consider the Galois group as a permutation of the roots of f. In the first case we must fix K, and in the second we must fix the larger field L. Thus we should expect $\Gamma_L(f)$ to be a subgroup of $\Gamma_K(f)$.

For example suppose that f is separable over K, that $F:K$ is a splitting field extension for f over K and that $F:L:K$. Then

$$\Gamma_L(f) = \Gamma(F:L) \subseteq \Gamma(F:K) = \Gamma_K(f).$$

In general L is not an intermediate field: the *theorem on natural irrationalities* says that in fact this does not affect things.

Theorem 11.9 *Suppose that $f \in K[x]$ and that $L:K$ is an extension. Let $N:L$ be a splitting field extension for f over L, let $\alpha_1, \ldots, \alpha_n$ be the roots of f in N and let $M = K(\alpha_1, \ldots, \alpha_n)$ (so that $M:K$ is a splitting field extension for f over K). Let L_0 be the fixed field of $\Gamma_L(f)$. Then if $\sigma \in \Gamma_L(f)$, $\sigma|_M \in \Gamma(M:L_0 \cap M)$,* and the map $\theta : \sigma \to \sigma|_M$ is an isomorphism of $\Gamma_L(f)$ onto the subgroup $\Gamma(M:L_0 \cap M)$ of $\Gamma_K(f)$.

Proof. We have the following diagram of inclusions.

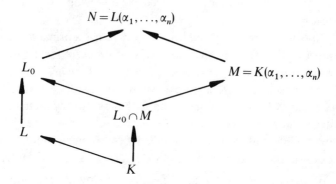

If $\sigma \in \Gamma(N:L)$, σ fixes K and permutes $\{\alpha_1, \ldots, \alpha_n\}$ and so $\sigma(M) \subseteq M$. Thus $\sigma|_M$ is an automorphism of M, which clearly fixes $L_0 \cap M$. Since the group multiplication is the composition of mappings, θ is a homomorphism of $\Gamma(N:L)$ into $\Gamma(M:L_0 \cap M)$.

If $\theta(\sigma)$ is the identity, σ fixes $\alpha_1, \ldots, \alpha_n$ and so (since σ fixes L) σ must be the identity on $N = L(\alpha_1, \ldots, \alpha_n)$. Thus θ is one–one.

Let V be the fixed field of $\theta(\Gamma(N:L))$. As we have seen, $V \supseteq L_0 \cap M$. Suppose that $x \in M$ and $x \notin L_0 \cap M$. Since L_0 is the fixed field of $\Gamma(N:L)$, there exists $\sigma \in \Gamma(N:L)$ such that $\sigma(x) \neq x$. Thus $\theta(\sigma)(x) \neq x$, and so $x \notin V$. Thus $V = L_0 \cap M$, and so, by Theorem 11.3

$$\theta(\Gamma(N:L)) = \Gamma(M:V) = \Gamma(M:L_0 \cap M).$$

Note that if f is separable over K, then f is separable over L, and so $L = L_0$: in this case the theorem becomes a little simpler.

12

Finite fields

In this chapter and the next we digress a little to consider some applications of the theory that we have developed so far.

12.1 A description of the finite fields

Suppose that K is a finite field: that is, a field with only finitely many elements. Then char $K = p > 0$, and we can identify $\mathbb{Z}_p$ and the prime subfield of K. As K is finite, $[K:\mathbb{Z}_p]$ must be finite. If $[K:\mathbb{Z}_p] = n$, then K is an n-dimensional vector space over $\mathbb{Z}_p$, and as a vector space, K is isomorphic to $(\mathbb{Z}_p)^n$. Thus $|K| = p^n$.

We shall now show that for each prime p and each positive integer n there is essentially just one field of order p^n.

Theorem 12.1 *For each prime number p and each positive integer n there is a field K with $|K| = p^n$. The field K is a splitting field for $f = x^{p^n} - x$ over its prime subfield. If K and K' are two fields of order p^n then K and K' are isomorphic.*

Proof. Let $K:\mathbb{Z}_p$ be a splitting field extension for f over $\mathbb{Z}_p$. Since $D(f) = -1$, f has p^n distinct roots in K, by Theorem 10.5. Let R denote the set of roots of f in K. Then

$$R = \{\alpha : \phi^n(\alpha) = \alpha\},$$

where ϕ is the Frobenius monomorphism. But $\{\alpha : \phi^n(\alpha) = \alpha\}$ is a subfield of K: thus R is a field, and f splits over R. Consequently $R = K$ and $|K| = |R| = p^n$.

Suppose now that L is a finite field of order p^n. $L^* = L \setminus \{0\}$ is a multiplicative group of order $p^n - 1$. As the order of an element of a group divides the order of the group, $\lambda^{p^n - 1} = 1$ for all λ in L^*. Bearing in mind that $0^{p^n} = 0$, this means that $\lambda^{p^n} = \lambda$ for all λ in L. Thus $f = x^{p^n} - x$ has p^n distinct roots in L, so that f splits over L and L is a splitting field for f over its prime subfield. By Corollary 1 of Theorem 7.5, K and L are isomorphic.

Corollary 1 *If K is a finite field, with prime subfield $\mathbb{Z}_p$, then $K:\mathbb{Z}_p$ is a Galois extension.*

Separability comes from the Corollary to Theorem 10.8. Using Theorem 10.1 and Corollary 3 to Theorem 9.1, we can strengthen this:

Corollary 2 *If $L:K$ is an extension and L is finite, then $L:K$ is a Galois extension.*

Exercises

12.1 Show that if K is a finite field of order q and p is a prime then there are exactly $(q^p - q)/p$ monic irreducible polynomials of degree p over K.

12.2 Show that if K is a finite field and n is a positive integer then there exists an (essentially unique) extension $L:K$ with $[L:K]=n$.

12.3 Suppose that $L:K$ is an extension and that L has p^n elements. Show that $|K|=p^d$, where $d|n$. Conversely, if $d|n$, show that $(x^{p^d}-x)|(x^{p^n}-x)$ and deduce that L has exactly one subfield with p^d elements.

12.2   An example

We have just seen that we can construct fields of all prime power orders p^n by constructing splitting field extensions for $f=x^{p^n}-x$ over $\mathbb{Z}_p$. The polynomial f is of course not irreducible; in certain circumstances we obtain more information by considering splitting fields of irreducible polynomials.

Let us illustrate this by considering fields of order p^p. Let us denote the elements of $\mathbb{Z}_p$ by $\bar{0}, \bar{1}, \ldots, \overline{p-1}$. We consider the polynomial

$$g=x^p-x-\bar{1}.$$

This has no roots in $\mathbb{Z}_p$ ($g(a)=-\bar{1}$ for all a in $\mathbb{Z}_p$); let $L:\mathbb{Z}_p$ be a splitting field extension for g over $\mathbb{Z}_p$, and let α be a root of g in L. Then if $\bar{j}\in\mathbb{Z}_p$,

$$(\alpha+\bar{j})^p-(\alpha+\bar{j})-\bar{1}=\alpha^p+\bar{j}-\alpha-\bar{j}-\bar{1}=\bar{0},$$

so that the roots of g are $\alpha, \alpha+\bar{1}, \ldots, \alpha+\overline{p-1}$. Note that it follows that $L=\mathbb{Z}_p(\alpha)$.

Next we show that g is irreducible over $\mathbb{Z}_p$. Suppose that $g=g_1g_2$, where g_1 and g_2 are monic, and $1\leqslant \text{degree } g_1=d<p$. Let

$$S=\{\bar{i}:\alpha+\bar{i} \text{ is a root of } g_1\}.$$

Then the coefficient of x^{d-1} in g_1 is

$$-\sum_{i \in S}(\alpha + \bar{i}) = -d\alpha + t,$$

where $t \in \mathbb{Z}_p$. As this coefficient is in $\mathbb{Z}_p$, $-d\alpha \in \mathbb{Z}_p$, and so $\alpha \in \mathbb{Z}_p$, giving a contradiction.

This means that $[L:\mathbb{Z}_p] = p$, and so $|L| = p^p$. As $L:\mathbb{Z}_p$ is also a splitting field extension for $x^{p^p} - x$, this means that $(x^p - x - \bar{1})|(x^{p^p} - x)$ in $\mathbb{Z}_p[x]$.

Notice also that, for each non-zero $\bar{j}$ in $\mathbb{Z}_p$, $\bar{j}\alpha, \bar{j}\alpha + \bar{1}, \dots, \bar{j}\alpha + p - 1$ are the roots of

$$x^p - x - \bar{j},$$

and an argument similar to that for g shows that this polynomial is also irreducible. Thus $L:\mathbb{Z}_p$ is also a splitting field extension for each of the polynomials $x^p - x - \bar{j}$ $(1 \leqslant j \leqslant p - 1)$.

Exercise

12.4 Factorize $x^{p^p} - x$ over $\mathbb{Z}_p$.

12.3 Some abelian group theory

In the next section, we shall investigate further the multiplicative group K^* of non-zero elements of a finite field. This group is abelian; we now study the structure of finite abelian groups.

Theorem 12.2 *Suppose that $(G, +)$ is a finite abelian group. G is isomorphic to a product of cyclic groups:*

$$G \cong \mathbb{Z}_{d_1} \times \cdots \times \mathbb{Z}_{d_s}.$$

Further the isomorphism can be chosen so that $d_j | d_k$ for $1 \leqslant j < k \leqslant s$. The number s is characterized by the property that G is generated by s elements, but it is not generated by $s - 1$ elements.

Proof. We prove this by induction on $|G|$. Suppose that the result is true for all abelian groups of order less than n, and that $|G| = n$.

There exists an integer s such that G is generated by s elements, but is not generated by fewer than s elements. Let m be the least positive number such that there exists a set $\{g_1, \dots, g_s\}$ of generators and a relation

$$mg_1 + a_2 g_2 + \cdots + a_s g_s = 0$$

(with $a_2, \dots, a_s$ in $\mathbb{Z}$). Note that $m > 1$, since otherwise G would be generated by $\{g_2, \dots, g_s\}$. We can write $a_i = mq_i + r_i$ with $0 \leqslant r_i < m$, for $2 \leqslant i \leqslant s$. Then if $h_1 = g_1 + q_2 g_2 + \cdots + q_s g_s$, G is generated by $\{h_1, g_2, \dots, g_s\}$ and

$$mh_1 + r_2 g_2 + \cdots + r_s g_s = 0.$$

The minimality of m implies that $r_2=r_3=\cdots=r_s=0$, and so $mh_1=0$. We now claim that G is isomorphic to $\langle h_1 \rangle \times \langle g_2,\ldots,g_s \rangle$. If $(a,b) \in \langle h_1 \rangle \times \langle g_2,\ldots,g_s \rangle$, let $\theta(a,b)=a+b$. The map θ is a homomorphism of $\langle h_1 \rangle \times \langle g_2,\ldots,g_s \rangle$ into G. It is an epimorphism, since $\{h_1,g_2,\ldots,g_s\}$ generates G. If (a,b) is in the kernel of θ, $a+b=0$. Writing $a=j_1h_1, b=j_2g_2+\cdots+j_rg_s$, with $0 \le j_1 < m$, we have

$$j_1h_1+j_2g_2+\cdots+j_sg_s=0.$$

It follows from the minimality of m that $j_1=0$, and so $a=b=0$. Thus θ is an isomorphism.

We now apply the inductive hypothesis to $\langle g_2,\ldots,g_s \rangle$, which is clearly generated by $s-1$ elements, but not by $s-2$ elements: the subgroup $\langle g_2,\ldots,g_s \rangle$ is isomorphic to

$$\mathbb{Z}_{d_2} \times \cdots \times \mathbb{Z}_{d_s},$$

with $d_j | d_k$ for $2 \le j < k \le s$. Consequently $G \cong \mathbb{Z}_m \times \mathbb{Z}_{d_2} \times \cdots \times \mathbb{Z}_{d_s}$. Let $h_1,\ldots,h_s$ be the corresponding generators in G. It follows from the minimality of m that $m \le d_2$. Let $d_2=e_2m+f_2$, where $0 \le f_2 < m$, and let $h_1'=h_1+e_2h_2$. Then G is generated by $\{h_1',h_2,\ldots,h_s\}$. As

$$mh_1'+f_2h_2=0$$

it follows that $f_2=0$, and so $m|d_2$. This completes the proof.

If G is a finite group, the *exponent* $e(G)$ of G is the least positive integer k such that $g^k=e$, for all g in G. We have already used the fact, in Theorem 12.1, that $g^{|G|}=e$ for all g in G. Thus $e(G) \le |G|$, and $e(G) \big| |G|$. If $g \in G$, we denote the order of g by $o(g)$: clearly $o(g) \le e(G)$, and $o(g)|e(G)$.

Some examples: in Σ_3, the elements have order 1, 2 or 3; the exponent of Σ_3 is 6. In Σ_6, the elements have order 1, 2, 3, 4, 5 or 6: $e(\Sigma_6)=60$, and $|\Sigma_6|=720$.

Corollary *Suppose that G is a finite abelian group. There exists g in G such that $o(g)=e(G)$.*
Proof. $G \cong \mathbb{Z}_{d_1} \times \cdots \times \mathbb{Z}_{d_s}$, with $d_j | d_k$ for $1 \le j < k \le s$. Then if $g \in G$, $g^{d_s}=e$; thus $e(G) \le d_s$. On the other hand, G has a subgroup isomorphic to $\mathbb{Z}_{d_s}$; if h is a generator of this, $o(h)=d_s$. As $o(h) \le e(G)$, this proves the result.

Exercises

12.5 Suppose that a and b are positive integers with highest common factor d. Show that

$$\mathbb{Z}_a \times \mathbb{Z}_b \cong \mathbb{Z}_d \times \mathbb{Z}_{ab/d}.$$

12.6 Show that a finite abelian group is isomorphic to a product of cyclic groups of prime power order.

12.7 Suppose that G is an abelian group. Show that the set T of elements of finite order is a subgroup of G and that every element of G/T, except the identity, is of infinite order.

12.8 Suppose that G is a finitely generated abelian group every element of which, except the identity, has infinite order. Show that $G \cong \mathbb{Z}^s$, where s is defined by the property that G is generated by s elements, but is not generated by $s-1$ elements.

12.9 Suppose that G is a finitely generated abelian group. Show that $G \cong \mathbb{Z}^s \times T$, where T is a finite group.

12.4 The multiplicative group of a finite field

Theorem 12.3 *Suppose that K is a field, with multiplicative group K^* of non-zero elements. If G is a finite subgroup of K^*, then G is cyclic.*
Proof. Let $\lambda = e(G)$. Then $\alpha^\lambda = 1$ for all $\alpha \in G$. As $x^\lambda - 1$ has at most λ roots, $|G| \leqslant e(G)$. But $e(G) \leqslant |G|$, so that $e(G) = |G|$. By Corollary to Theorem 12.2, G has an element of order $|G|$, and so G is cyclic.

Corollary 1 *If K is a finite field, K^* is cyclic.*

Corollary 2 *If $L:K$ is an extension, and L is a finite field, then $L:K$ is simple.*
Proof. Let α generate the multiplicative group L^*. Then $L = K(\alpha)$.

12.5 The automorphism group of a finite field

Theorem 12.4 *Suppose that K is a finite field with p^n elements. Then the group of all automorphisms of K is cyclic of order n, and is generated by the Frobenius automorphism ϕ.*
Proof. We can identify the prime subfield of K with $\mathbb{Z}_p$. Every automorphism of K fixes $\mathbb{Z}_p$. As $[K:\mathbb{Z}_p] = n$ and $K:\mathbb{Z}_p$ is a Galois extension, there are exactly n automorphisms of K, by Theorem 10.4. Let d denote the order of ϕ. Then

$$\alpha^{p^d} = \phi^d(\alpha) = \alpha \text{ for each } \alpha \text{ in } K$$

so that the polynomial $x^{p^d} - x$ has p^n roots in K. This implies that $d \geqslant n$. As $d | n$, we must have that $d = n$, and that ϕ generates the group of automorphisms of K.

Corollary *Suppose that $L:K$ is an extension and that L is finite. Then $\Gamma(L:K)$ is cyclic of order $[L:K]$.*
Proof. Suppose that L has characteristic p. As $K:\mathbb{Z}_p$ is a Galois extension, $\Gamma(L:K) \cong \Gamma(L:\mathbb{Z}_p)/\Gamma(K:\mathbb{Z}_p)$ by the fundamental theorem of Galois theory,

and so $\Gamma(L:K)$ is cyclic. Also $L:K$ is a Galois extension, and so $|\Gamma(L:K)| = [L:K]$.

Exercises

12.10 Use Theorem 12.4 to give another solution to Exercise 12.3.

12.11 Suppose that p and q are primes and that $p < q$. Show that if p does not divide $q - 1$ then there is an extension $L:\mathbb{Z}_q$ which is a splitting field extension for each of the polynomials $x^p - a$ (a a non-zero element of $\mathbb{Z}_q$).

13

The theorem of the primitive element

In this chapter, we consider the problem: if $L:K$ is algebraic, under what circumstances is L a simple extension of K?

13.1 A criterion in terms of intermediate fields

Theorem 13.1 *An algebraic extension $L:K$ is simple if and only if there are only finitely many intermediate fields.*

Proof. First suppose that there are only finitely many intermediate fields. $L:K$ must be finitely generated over K, for otherwise there is a strictly increasing infinite sequence of intermediate fields. Thus $L:K$ is finite (Theorem 4.6). If K is finite, L is finite, and so $L:K$ is simple (Corollary 2 of Theorem 12.3). We may therefore restrict attention to the case where K is *infinite*.

As we have observed, L is finitely generated over K. Let

$$r = \inf\{|A| : L = K(A)\}.$$

We want to show that $r = 1$. Suppose on the contrary that $r \geqslant 2$ and that $L = K(\alpha_1, \alpha_2, \ldots, \alpha_r)$. Let $M = K(\alpha_1, \alpha_2)$. For each β in K, let

$$F_\beta = K(\alpha_1 + \beta\alpha_2).$$

As K is infinite, and as there are only finitely many fields intermediate between K and L, there exist β and γ in K, with $\beta \neq \gamma$, such that $F_\beta = F_\gamma$. But then

$$(\alpha_1 + \beta\alpha_2) - (\alpha_1 + \gamma\alpha_2) = (\beta - \gamma)\alpha_2 \in F_\beta,$$

and so $\alpha_2 \in F_\beta$. Also $\alpha_1 = (\alpha_1 + \beta\alpha_2) - \beta\alpha_2 \in F_\beta$, so that $K(\alpha_1, \alpha_2) \subseteq K(\alpha_1 + \beta\alpha_2)$. Consequently

$$L = K(\alpha_1 + \beta\alpha_2, \alpha_3, \ldots, \alpha_r)$$

contradicting the minimality of r.

Conversely suppose that $L = K(\alpha)$ is simple and algebraic over K. Let m be the minimal polynomial for α over K. m is irreducible over K, but of course m factorizes over L. Nevertheless, m has only finitely many monic divisors $d_1, \ldots, d_k$, say, in $L[x]$. Now suppose that F is an intermediate field. Let m_F be the minimal polynomial for α over F. Considering m as an element of $F[x]$, we see that $m_F | m$. But this means that $m_F | m$ in $L[x]$, and so $m_F = d_i$ for some $1 \leqslant i \leqslant k$. The proof will therefore be complete if we can show that m_F determines F. Let

$$m_F = a_0 + a_1 x + \cdots + a_r x^r,$$

and let $F_0 = K(a_0, \ldots, a_r)$. Then $F_0 \subseteq F$, and so m_F is irreducible over F_0. Thus m_F is the minimal polynomial for α over F_0. As $L = F_0(\alpha)$, $[L:F_0] =$ degree m_F (Theorem 4.4). But $[L:F] =$ degree m_F, by the same argument, so that as $F \supseteq F_0$ we must have $F = F_0 = K(a_0, \ldots, a_r)$. Thus m_F determines F; this completes the proof.

Exercise

13.1 Use Exercise 1.17 to give another proof that $L:K$ is simple if K is infinite and there are only finitely many intermediate fields.

13.2 Suppose that $K(t):K$ is a simple transcendental extension. Show that there are infinitely many intermediate fields.

13.2   The theorem of the primitive element
This rather quaint title is given to the following theorem.

Theorem 13.2 *Suppose that $L:K$ is finite and separable. Then $L:K$ is simple.*
Proof. Suppose that $\alpha_1, \ldots, \alpha_n$ generate L over K. Let $g = m_{\alpha_1} \ldots m_{\alpha_n}$, where m_{α_i} is the minimal polynomial of α_i over K. Then g is separable over K. Let $N:L$ be a splitting field extension for g over L. As $\alpha_1, \ldots, \alpha_n$ are roots of g, $N:K$ is also a splitting field extension for g over K. Thus $N:K$ is normal (Corollary 1 to Theorem 9.1) and separable (Corollary 3 to Theorem 10.3) and is therefore a Galois extension. Thus K is the fixed field of $\Gamma(N:K)$.

Now $\Gamma(N:K)$ is finite, and so it has finitely many subgroups. By the fundamental theorem of Galois theory, these are in one–one correspondence with the fields intermediate between N and K. Thus there are finitely many fields intermediate between N and K, and *a fortiori* there are finitely many fields intermediate between L and K. The result follows from Theorem 13.1.

Corollary *If $L:K$ is a Galois extension, there exists an irreducible polynomial f in $K[x]$ such that $L:K$ is a splitting field extension for f over K.*

Exercises

13.3 Let p be a prime, let $J = \mathbb{Z}_p(\alpha)$, where α is transcendental over $\mathbb{Z}_p$, and let $K = J(\beta)$, where β is transcendental over J. Let $L:K$ be a splitting field extension for $(x^p - \alpha)(x^p - \beta)$.

   (i) Show that $[L:K] = p^2$.

   (ii) Show that if $\gamma \in L$ then $\gamma^p \in K$.

   (iii) Show that $L:K$ is not simple.

   (iv) In the case where $p = 2$, find all the intermediate fields $L:M:K$.

13.4 Suppose that $L:K$ is a Galois extension with Galois group $\{\sigma_1, \ldots, \sigma_n\}$ and that $\alpha \in L$. Show that $L = K(\alpha)$ if and only if $(\sigma_1(\alpha), \ldots, \sigma_n(\alpha))$ is a basis for L over K.

13.5 Suppose that $L:K$ is a finite separable extension and that $M:L$ is a finite simple extension. Show that $M:K$ is a simple extension.

13.3   An example

Let us consider a very easy example. $\mathbb{Q}(\sqrt{2}, \sqrt{3}):\mathbb{Q}$ is a splitting field extension for $f = x^4 - 5x^2 + 6 = (x^2 - 2)(x^2 - 3)$ over $\mathbb{Q}$. $[\mathbb{Q}(\sqrt{2}, \sqrt{3}):\mathbb{Q}] = 4$, and the Galois group $\Gamma_{\mathbb{Q}}(f)$ is best described by its action on $\sqrt{2}$ and $\sqrt{3}$:

	$\sigma_0 = e$	σ_1	σ_2	σ_3
$\sqrt{2}$	$\sqrt{2}$	$-\sqrt{2}$	$\sqrt{2}$	$-\sqrt{2}$
$\sqrt{3}$	$\sqrt{3}$	$\sqrt{3}$	$-\sqrt{3}$	$-\sqrt{3}.$

$\Gamma_{\mathbb{Q}}(f)$ is isomorphic to $\mathbb{Z}_2 \times \mathbb{Z}_2$ and has three non-trivial subgroups: $\{\sigma_0, \sigma_1\}$, $\{\sigma_0, \sigma_2\}$ and $\{\sigma_0, \sigma_3\}$. The corresponding fixed fields are $\mathbb{Q}(\sqrt{3})$, $\mathbb{Q}(\sqrt{2})$ and $\mathbb{Q}(\sqrt{6})$. If α is any element of $\mathbb{Q}(\sqrt{2}, \sqrt{3})$ which does not belong to any of these three intermediate fields, then $\mathbb{Q}(\sqrt{2}, \sqrt{3}) = \mathbb{Q}(\alpha)$.

14

Cubics and quartics

In this chapter we shall see how the theory that we have developed so far relates to the solution of cubic and quartic equations. In the process, we shall introduce some ideas which will appear again when we return to the general theory.

14.1　Extension by radicals

We have seen that we can deal with quadratic polynomials by constructing a splitting field by adjoining a square root. Having done this, we have a procedure for factorizing the polynomial.

In this chapter we shall see that similar results hold for cubics and quartics. In order to see what we are trying to achieve, let us make some definitions.

If $L:K$ is an extension, and $\beta \in L$, we say that β is a *radical* over K if $\beta^n \in K$ for some n. Thus a radical over K is an nth root of some element of K, possibly in a larger field.

We shall say that an extension $L:K$ is an *extension by radicals* if there are intermediate fields

$$L = L_r : L_{r-1} : \ldots : L_0 = K$$

such that $L_i = L_{i-1}(\beta_i)$, with β_i a radical over L_{i-1}, for $1 \leqslant i \leqslant r$. Thus $L:K$ is an extension by radicals if L can be obtained by successively adjoining radicals.

Now suppose that $f \in K[x]$. We say that f is *solvable by radicals* if there is an extension $L:K$ by radicals such that f splits over L. It is important to note that L need not be a splitting field for f – it may be considerably larger.

The general problems that arise, then, and that we shall consider in the subsequent chapters, are to determine whether $f \in K[x]$ is solvable by radicals or not, and, if so, to find a procedure for factorizing f.

14.2 The discriminant

Suppose that f is a separable irreducible monic cubic polynomial in $K[x]$. Then the Galois group $\Gamma_K(f)$ acts transitively on the three roots of f in a splitting field, and so it must be either the full permutation group Σ_3 or the alternating group A_3. How can we determine which it is?

This is a problem which we can consider quite generally. We shall, however, suppose that char $K \neq 2$. Suppose that f is a polynomial in $K[x]$ and that $\alpha_1, \ldots, \alpha_n$ are roots of f (repeated according to multiplicity) in a splitting field extension $L:K$. We set

$$\delta = \prod_{1 \leqslant i < j \leqslant n} (\alpha_j - \alpha_i)$$

If f has a repeated root then $\delta = 0$; otherwise, f is separable, and $\delta \neq 0$. If $\sigma \in \Gamma_K(f)$ then

$$\sigma(\delta) = \prod_{1 \leqslant i < j \leqslant n} (\sigma(\alpha_j) - \sigma(\alpha_i)) = \varepsilon_\sigma \delta,$$

where $\varepsilon_\sigma = 1$ if σ is an even permutation of $\alpha_1, \ldots, \alpha_n$ and $\varepsilon_\sigma = -1$ if σ is an odd permutation.

There are therefore three possibilities. First, $\delta = 0$: in this case f has a repeated root. Secondly, δ is a non-zero element of K. In this case, δ is in the fixed field of $\Gamma_K(f)$, so that $\Gamma_K(f) \subseteq A_n$. Thirdly, $\delta \notin K$. In this case, δ is not in the fixed field of $\Gamma_K(f)$, so that $\Gamma_K(f) \nsubseteq A_n$. On the other hand $\Delta = \delta^2$ is fixed by $\Gamma_K(f)$, so that $x^2 - \Delta$ is the minimal polynomial of δ, and $[K(\delta):K] = 2$. Now $\Gamma_K(f) \cap A_n$ has index 2 in $\Gamma_K(f)$, so that it follows from the fundamental theorem of Galois theory that $K(\delta)$ is the fixed field of $\Gamma_K(f) \cap A_n$, and $\Gamma_K(f) \cap A_n = \Gamma(L:K(\delta))$.

The quantity $\Delta = \delta^2$ is called the *discriminant* of f. Notice that, although δ depends on the order in which we label the roots of f, Δ does not. Let us sum up our discussion in terms of Δ.

Theorem 14.1 Suppose that char $K \neq 2$ and that $f \in K[x]$. Let Δ be the discriminant of $f \in K[x]$, and let $L:K$ be a splitting field extension for f.

(i) If $\Delta = 0$, f has a repeated root in L.

(ii) If $\Delta \neq 0$ and Δ has a square root in K, then $\Gamma_K(f) \subseteq A_n$.

(iii) If Δ has no square root in K, it has a square root δ in L. $\Gamma_K(f) \nsubseteq A_n$, and $K(\delta)$ is the fixed field of $\Gamma_K(f) \cap A_n$.

In practice, it is not hard to calculate the discriminant. The quantity δ is given by the Vandermonde determinant:

$$\delta = \begin{vmatrix} 1 & 1 & \dots & 1 \\ \alpha_1 & \alpha_2 & \dots & \alpha_n \\ \vdots & \vdots & \ddots & \vdots \\ \alpha_1^{n-1} & \alpha_2^{n-1} & \dots & \alpha_n^{n-1} \end{vmatrix}.$$

If we multiply the matrix by its transpose and evaluate the determinant we find that

$$\Delta = \begin{vmatrix} n & \lambda_1 & \dots & \lambda_{n-1} \\ \lambda_1 & \lambda_2 & \dots & \lambda_n \\ \vdots & \vdots & \ddots & \vdots \\ \lambda_{n-1} & \lambda_n & \dots & \lambda_{2n-2} \end{vmatrix},$$

where $\lambda_j = \alpha_1^j + \dots + \alpha_n^j$. The quantities λ_j can be expressed in terms of the coefficients of f (as we shall see in Chapter 19), and so we can calculate Δ.

For example, if

$$f = x^2 + a_1 x + a_0,$$

then

$$\Delta = a_1^2 - 4a_0, \text{ while if}$$
$$f = x^3 + a_2 x^2 + a_1 x + a_0,$$

then

$$\Delta = -4a_2^3 a_0 + a_2^2 a_1^2 + 18a_2 a_1 a_0 - 4a_1^3 - 27a_0^2.$$

Exercises

14.1 Suppose that f is a polynomial in $K[x]$, with roots $\alpha_1, \dots, \alpha_n$ in some splitting field extension. Show that

$$\Delta = \eta_n \prod_{j=1}^{n} Df(\alpha_j),$$

where $\eta_n = 1$ if $n \pmod 4 = 0$ or 1 and $\eta_n = -1$ otherwise.

14.2 Suppose that

$$f = a_0 + a_1 x + \dots + a_n x^n$$

is a polynomial of degree n in $K[x]$ and that $\alpha_1, \dots, \alpha_n$ are roots of f in a splitting field L.

(i) Show that, in $L[x]$, $f = (x - \alpha_i)g_i$, where

$$g_i = a_1 + a_2(x + \alpha_i) + \dots + a_n(x^{n-1} + \alpha_i x^{n-2} + \dots + \alpha_i^{n-1}).$$

(ii) Show that $Df = \sum_{i=1}^{n} g_i$.

(iii) Let $\lambda_j = \sum_{i=1}^{n} \alpha_i^j$, for $j = 1, 2, \ldots$ Establish *Newton's identities*:

$$a_{n-1} + a_n \lambda_1 \qquad\qquad = 0,$$

$$2a_{n-2} + a_{n-1}\lambda_1 + a_n\lambda_2 \qquad = 0,$$

$$\vdots$$

$$na_0 + a_1\lambda_1 + \cdots + a_{n-1}\lambda_{n-1} + a_n\lambda_n \qquad = 0,$$

and

$$a_0\lambda_k + a_1\lambda_{k+1} + \cdots + a_{n-1}\lambda_{k+n-1} + a_n\lambda_{k+n} = 0$$

for $k = 1, 2, 3, \ldots$

14.3 Suppose that $f = x^n + px + q$. Show that

$$\lambda_1 = \lambda_2 = \cdots = \lambda_{n-2} = 0,$$

$$\lambda_{n-1} = -(n-1)p,$$

$$\lambda_n = -nq,$$

$$\lambda_{n+1} = \cdots = \lambda_{2n-3} = 0$$

and

$$\lambda_{2n-2} = (n-1)p^2.$$

Show that the discriminant Δ of f is

$$\Delta = \eta_{n+1} n^n q^{n-1} - \eta_n (n-1)^{n-1} p^n$$

where $\eta_n = 1$ if $n \pmod 4 = 0$ or 1 and $\eta_n = -1$ otherwise.

14.4 Suppose that char $K = 2$ and that $f \in K[x]$ is separable. Show that the discriminant of f always has a square root in K. Give an example to show that Theorem 14.1 does not hold for fields of characteristic 2.

14.3   Cubic polynomials

Suppose again that f is an irreducible monic cubic polynomial in $K[x]$:

$$f = x^3 + a_2 x^2 + a_1 x + a_0.$$

In order to simplify things, we shall assume that char K is not equal to 2 or 3. In particular, this means that f is separable. We can simplify the expression for f by setting $y = x + a_2/3$: then

$$f = y^3 + py + q,$$

where $p = a_1 - a_2^2/3$ and $q = a_0 + 2a_2^3/27 - a_2a_1/3$. We therefore consider the polynomial

$$g = x^3 + px + q.$$

Let $L:K$ be a splitting field extension for g over K, and let $\alpha_1, \alpha_2, \alpha_3$ be the roots of g in L. g has discriminant $\Delta = -4p^3 - 27q^2$: let δ be a square root of Δ in L. Then we know, by Theorem 14.1, that $[L:K(\delta)] = 3$ and that $\Gamma(L:K(\delta))$ is the cyclic group A_3.

How do we proceed from here? Our aim is to solve g by radicals: this suggests that we should adjoin a cube root of a suitable element θ of K or $K(\delta)$. In fact, it is convenient to proceed rather indirectly. If we have such an element θ then in a suitable splitting field $x^3 - \theta$ factorizes as

$$x^3 - \theta = (x - \beta)(x - \omega\beta)(x - \omega^2\beta)$$

where ω and ω^2 are cube roots of unity. As a next step, then, we adjoin cube roots of unity: let $L(\omega):L$ be a splitting field extension for $x^3 - 1 = (x - 1)(x^2 + x + 1)$. We now have the following diagram of inclusions.

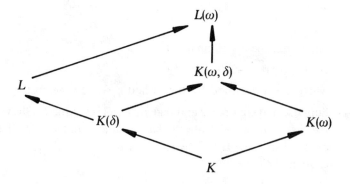

Note that there are essentially only two possibilities. First it may happen that $x^3 - 1$ splits over $K(\delta)$. In this case $L = L(\omega)$ and we do not need to extend further. Secondly it may happen that $x^3 - 1$ does not split over $K(\delta)$. In this case $[K(\omega, \delta):K(\delta)] = 2$, while $[L:K(\delta)] = 3$, and so $x^3 - 1$ does not split over $L:L \neq L(\omega)$, and we go beyond a splitting field for g.

We now return to the cubic g. In $L(\omega)$ we set

$$\beta = \alpha_1 + \omega\alpha_2 + \omega^2\alpha_3$$
$$\gamma = \alpha_1 + \omega^2\alpha_2 + \omega\alpha_3.$$

Then

$$\beta\gamma = \alpha_1^2 + \alpha_2^2 + \alpha_3^2 + (\omega + \omega^2)(\alpha_1\alpha_2 + \alpha_1\alpha_3 + \alpha_2\alpha_3)$$
$$= (\alpha_1 + \alpha_2 + \alpha_3)^2 - 3(\alpha_1\alpha_2 + \alpha_1\alpha_3 + \alpha_2\alpha_3) = -3p,$$

so that $\beta^3\gamma^3 = -27p^3$; and

$$\beta^3 + \gamma^3 = (\alpha_1 + \omega\alpha_2 + \omega^2\alpha_3)^3 + (\alpha_1 + \omega^2\alpha_2 + \omega\alpha_3)^3 + (\alpha_1 + \alpha_2 + \alpha_3)^3$$
$$= 3(\alpha_1^3 + \alpha_2^3 + \alpha_3^3) + 18\alpha_1\alpha_2\alpha_3 = -27q.$$

Thus

$$(x - \beta^3)(x - \gamma^3) = x^2 + 27qx - 27p^3,$$

and so β^3 and γ^3 are the elements

$$-\tfrac{27}{2}q \pm \tfrac{3}{2}\sqrt{-3\Delta} = \tfrac{27}{2}q \pm \tfrac{3}{2}(2\omega + 1)\delta.$$

Consequently we can obtain β^3 and γ^3 by adjoining a square root of -3Δ to K, and then obtain β by adjoining a cube root: then $\gamma = -3p/\beta$. Finally

$$\alpha_1 = \tfrac{1}{3}(\beta + \gamma)$$
$$\alpha_2 = \tfrac{1}{3}(\omega^2\beta + \omega\gamma)$$
$$\alpha_3 = \tfrac{1}{3}(\omega\beta + \omega^2\gamma).$$

Notice that in these calculations we are essentially working in $L(\omega)$; as we have observed, this may well be larger than the splitting field L.

Exercise

14.5 Suppose that char K is not 2 or 3, and that $f = x^3 + px + q \in K[x]$. Let α be a root of f in a splitting field, let $g = 3x^2 - 3\alpha x - p$, and let β be a root of g in a splitting field for g over $K(\alpha)$. Express α in terms of β and show that β is a root of

$$h = 27x^6 + 27q^3 - p^3 \in K[x].$$

Conclude that $\alpha = \beta - p/3\beta$, where $\beta^3 = -q/2 + \delta$ and $\delta^2 = q^2/4 + p^3/27$: the cubic f can be solved by extracting a square root and a cube root.

14.4 Quartic polynomials

Suppose now that f is an irreducible monic quartic in $K[x]$:

$$f = x^4 + a_3x^3 + a_2x^2 + a_1x + a_0.$$

We continue to suppose that char K is not equal to 2 or 3. If we write $y = x + a_3/4$, f has the form

$$g = y^4 + py^2 + qy + r.$$

We therefore consider a polynomial of the form

$$g = x^4 + px^2 + qx + r.$$

Let $L:K$ be a splitting field extension for g over K, and let $\alpha_1, \alpha_2, \alpha_3$ and α_4 be the roots of g in L.

Let G be the Galois group $\Gamma_K(g)$. G can be considered, by its action on the roots of g, as a transitive subgroup of Σ_4. Now the Viergruppe N is a normal subgroup of Σ_4, and so $H = N \cap G$ is a normal subgroup of G. Let M be the fixed field of H. Then by the fundamental theorem of Galois theory $\Gamma(L:M) = H$ and $\Gamma(M:K) \cong G/H$.

Now H is an abelian group of order 1, 2 or 4 (in fact the first possibility cannot arise) and, as H is the kernel of the homomorphism ϕi, where i is the inclusion map $G \to \Sigma_4$ and ϕ is the epimorphism of Σ_4 onto Σ_3 described in Chapter 1, G/H is isomorphic to a subgroup of Σ_3, by the first isomorphism theorem for groups.

This suggests that we should first attempt to determine the intermediate field M. Let

$$\beta = \alpha_1 + \alpha_2, \quad \gamma = \alpha_1 + \alpha_3 \quad \text{and} \quad \delta = \alpha_1 + \alpha_4.$$

Then

$$\beta^2 = (\alpha_1 + \alpha_2)^2 = -(\alpha_1 + \alpha_2)(\alpha_3 + \alpha_4),$$
$$\gamma^2 = (\alpha_1 + \alpha_3)^2 = -(\alpha_1 + \alpha_3)(\alpha_2 + \alpha_4),$$

and

$$\delta^2 = (\alpha_1 + \alpha_4)^2 = -(\alpha_1 + \alpha_4)(\alpha_2 + \alpha_3).$$

Consequently β^2, γ^2 and δ^2 are in M, and so $K(\beta^2, \gamma^2, \delta^2) \subseteq M$. On the other hand, if σ is a permutation of $\alpha_1, \alpha_2, \alpha_3$ and α_4 which fixes β^2, γ^2 and δ^2, then $\sigma \in N$. Thus

$$\Gamma(L:K(\beta^2, \gamma^2, \delta^2)) \subseteq H = \Gamma(L:M)$$

and so $K(\beta^2, \gamma^2, \delta^2) \supseteq M$. Thus $M = K(\beta^2, \gamma^2, \delta^2)$.

Easy but tedious calculations show that

$$\beta^2 + \gamma^2 + \delta^2 = -2p,$$
$$\beta^2\gamma^2 + \beta^2\delta^2 + \gamma^2\delta^2 = p^2 - 4r,$$

and

$$\beta\gamma\delta = -q;$$

thus $K(\beta^2, \gamma^2, \delta^2):K$ is a splitting field extension for

$$x^3 + 2px^2 + (p^2 - 4r)x - q^2.$$

This cubic is called the *cubic resolvent* for g. By the results of the previous section, we can construct β^2, γ^2 and δ^2 by adjoining square roots and cube roots; we can then construct β, γ and δ by adjoining square roots (note, though, that $\beta\gamma\delta = -q$, so that some care is needed in the choice of signs). Then

$$\alpha_1 = \tfrac{1}{2}(\beta + \gamma + \delta),$$
$$\alpha_2 = \tfrac{1}{2}(\beta - \gamma - \delta),$$
$$\alpha_3 = \tfrac{1}{2}(-\beta + \gamma - \delta),$$

and

$$\alpha_4 = \tfrac{1}{2}(-\beta - \gamma + \delta).$$

Notice that this means that $L = K(\beta, \gamma, \delta)$.

What are the possible Galois groups of an irreducible quartic? The exercises which follow provide an answer to this question.

Exercises

14.6 Suppose that G is a transitive subgroup of Σ_4. Show that G is either (i) Σ_4, (ii) A_4, (iii) the Viergruppe N, (iv) cyclic of order 4 or (v) a non-abelian group of order 8, isomorphic to the group of rotations and reflections of a square.

14.7 Suppose that f is an irreducible quartic in $K[x]$ (where char K is not 2 or 3) and that $L:K$ is a splitting field extension for f. Let g be the cubic resolvent for f, and let M be a splitting field for g in L. Verify that the following table includes all possibilities and that it determines the Galois group of f in each case.

Discriminant	g	f	$\Gamma_K(f)$
No square root in K	Irreducible over K		Σ_4
Has square root in K	Irreducible over K		A_4
Has square root in K	Factorizes in $K[x]$		Viergruppe
No square root in K	Factorizes in $K[x]$	Factorizes in $M[x]$	Cyclic of order 4
No square root in K	Factorizes in $K[x]$	Irreducible over M	Of order 8

14.8 Determine the Galois groups of the following quartics in $\mathbb{Q}[x]$:

(i) $x^4 + 4x + 2$;

(ii) $x^4 + 8x - 12$;

(iii) $x^4 + 1$;

(iv) $x^4 + x^3 + x^2 + x + 1$;

(v) $x^4 - 2$.

15

Roots of unity

We have seen that in order to deal with cubic polynomials it is helpful to have cube roots of unity at our disposal. In this chapter we shall consider splitting fields and Galois groups of polynomials of the form $x^m - 1$ over a field K.

Technical problems can arise if char $K \neq 0$. Suppose that char $K = p > 0$ and that $m = p^r q$, where p does not divide q. Then in $K[x]$,

$$x^m - 1 = (x^q - 1)^{p^r};$$

thus a splitting field extension for $x^q - 1$ is a splitting field extension for $x^m - 1$: we need only consider the polynomial $x^q - 1$. For this reason, *in this chapter we shall suppose that* char K *does not divide* m. In this case, $D(x^m - 1) = mx^{m-1} \neq 0$, and so $x^m - 1$ has m distinct roots in a splitting field.

15.1 Cyclotomic polynomials

Suppose that $L:K$ is a splitting field extension for $x^m - 1$ over K. As $x^m - 1$ has m distinct roots, $L:K$ is a Galois extension. The set R of roots in L clearly forms a group under multiplication, and so, by Theorem 12.3, R is a cyclic group of order m. An element ε of R is called a *primitive mth root of unity* if ε generates R. Thus an element ε of L is a primitive mth root of unity if and ony if $\varepsilon^m = 1$ and $\varepsilon^j \neq 1$ for $1 \leqslant j < m$. For example, in $\mathbb{C}$, i and $-i$ are the primitive fourth roots of unity: -1 is the only primitive second root of unity and 1 is the only first root of unity. Notice that if ε is a primitive mth root of unity then $L = K(\varepsilon)$.

We now define the *mth cyclotomic polynomial* Φ_m to be

$$\Phi_m = \prod_\varepsilon (x - \varepsilon)$$

where the product is taken over all *primitive mth* roots of unity. An element

α in L is a root of $x^m - 1$ if and only if it is a primitive dth root of unity for some d which divides m: thus

$$x^m - 1 = \prod_{d|m} \Phi_d.$$

For example, in $\mathbb{C}[x]$

$$\Phi_1 = x - 1, \quad \Phi_3 = (x - \omega)(x - \omega^2) = x^2 + x + 1$$
$$\Phi_2 = x + 1, \quad \Phi_4 = (x - i)(x + i) = x^2 + 1$$

and

$$x^4 - 1 = (x - 1)(x + 1)(x^2 + 1) = \Phi_1 \Phi_2 \Phi_4.$$

We have defined Φ_m as an element of $L[x]$. In fact, as the examples suggest, we can say much more.

Theorem 15.1 $\Phi_m \in K_0[x]$, where K_0 is the prime subfield of K. If $K_0 = \mathbb{Q}$ then $\Phi_m \in \mathbb{Z}[x]$.

Proof. Since $x^m - 1 = \prod_{d|m} \Phi_d$, the theorem follows from an inductive application of the following elementary lemma.

Lemma 15.2 (i) *If $L:K$ is an extension, if $q \in L[x]$ and if there exist non-zero f and g in $K[x]$ such that $f = qg$, then $q \in K[x]$.*

(ii) *Suppose that K is the field of fractions of an integral domain R, that $q \in K[x]$ and that there exist monic f and g in $R[x]$ such that $f = qg$. Then $q \in R[x]$.*

Proof. This is just a matter of long division.

(i) Let

$$q = a_0 + a_1 x + \cdots + a_m x^m,$$
$$g = b_0 + b_1 x + \cdots + b_n x^n,$$
$$f = c_0 + c_1 x + \cdots + c_{m+n} x^{m+n}$$

where $a_m \neq 0, b_n \neq 0, c_{m+n} \neq 0$. As $a_m b_n = c_{m+n}$, $a_m \in K$. Suppose that we have shown that $a_i \in K$ for $i > j$. Then as

$$a_j b_n + a_{j+1} b_{n-1} + \cdots + a_m b_{n+j-m} = c_{n+j}$$

(where we set $b_k = 0$ if $k < 0$), $a_j \in K$.

(ii) In this case $b_n = 1$, and the same induction goes through.

Exercises

15.1 Show that the degree of Φ_m is $m \prod_{p|m} ((p - 1)/p)$, where the product is taken over all primes p which divide m.

15.2 Show that if n is odd then $\Phi_{2n}(x) = \Phi_n(-x)$.

15.3 Show that if p is a prime then
$$\Phi_{p^n}(x) = 1 + x^{p^{n-1}} + x^{2p^{n-1}} + \cdots + x^{(p-1)p^{n-1}}$$

15.2   Irreducibility

By Theorem 15.1, we can consider the cyclotomic polynomials Φ_m as polynomials in $K_0[x]$, where K_0 is the prime subfield of K. In the case where char $K_0 \neq 0$, the irreducibility of Φ_m over K_0 depends upon m and char K_0: for example, $\Phi_3 = x^2 + x + \bar{1}$ is irreducible over $\mathbb{Z}_5$, while over $\mathbb{Z}_7$

$$x^2 + x + \bar{1} = (x - \bar{2})(x - \bar{4}).$$

(The irreducibility of cyclotomic polynomials over finite fields is the subject of Exercises 15.6–15.10.)

In the important case where char $K_0 = 0$, the result is simple to state, but remarkably difficult to prove:

Theorem 15.3 *For each* m, Φ_m *is irreducible over* $\mathbb{Q}$.
Proof. Suppose that Φ_m is not irreducible. By Gauss' lemma we can write $\Phi_m = fg$, where f and g are in $\mathbb{Z}[x]$ and f is an irreducible monic polynomial with $1 \leqslant$ degree $f <$ degree Φ_m.

Let $L : \mathbb{Q}$ be a splitting field extension for Φ_m over $\mathbb{Q}$. We shall first show that, if ε is a root of f in L and p is a prime which does not divide m, then ε^p is a root of f.

Suppose not. Then, as ε^p is a primitive mth root of unity, $g(\varepsilon^p) = 0$. We define k in $\mathbb{Z}[x]$ by setting $k(x) = g(x^p)$. Then $k(\varepsilon) = g(\varepsilon^p) = 0$. Since f is the minimal polynomial for ε over $\mathbb{Q}$, $f \mid k$ in $\mathbb{Q}[x]$, and, by Lemma 15.2, we can write $k = fh$, with h in $\mathbb{Z}[x]$.

We now consider the quotient map: $n \to \bar{n}$ from $\mathbb{Z}$ onto $\mathbb{Z}_p$, and the induced map: $j \to \bar{j}$ of $\mathbb{Z}[x]$ onto $\mathbb{Z}_p[x]$. Under this map, $\bar{f}\bar{h} = \bar{k}$. But

$$\overline{k(x)} = \overline{g(x^p)} = (\overline{g(x)})^p$$

and so $\bar{f}\bar{h} = (\bar{g})^p$. Let $\bar{q}$ be any irreducible factor of $\bar{f}$ in $\mathbb{Z}_p[x]$. Then $\bar{q} \mid (\bar{g})^p$, and so $\bar{q} \mid \bar{g}$. This means that $\bar{q}^2 \mid \bar{f}\bar{g}$, so that $\bar{\Phi}_m = \bar{f}\bar{g}$ has a repeated root in a splitting field extension over $\mathbb{Z}_p$. But we have seen that this is not so, since p does not divide m.

Now let η be a root of f, and let θ be a root of g. θ and η are both primitive mth roots of unity, and so there exists r such that $\theta = \eta^r$, where r and m are relatively prime. We can write $r = p_1 \ldots p_k$ as a product of primes, where no p_i divides m. Repeated application of the result that we have proved shows that θ is a root of f. This means that Φ_m has a repeated root in L, and we know that this is not so.

Exercise

15.4 Suppose that ε is a primitive mth root of unity over $\mathbb{Q}$, where $m > 2$. Let $\eta = \varepsilon + \varepsilon^{-1}$. Show that $[\mathbb{Q}(\varepsilon):\mathbb{Q}(\eta)] = 2$, find the minimal polynomial for ε over $\mathbb{Q}(\eta)$ and identify the Galois group $\Gamma[\mathbb{Q}(\varepsilon):\mathbb{Q}(\eta)]$.

15.3 The Galois group of a cyclotomic polynomial

Suppose that $L:K$ is a splitting field extension for the cyclotomic polynomial Φ_m over K. If ε is a primitive mth root of unity then, as we have seen, $L = K(\varepsilon)$.

We can write the primitive mth roots of unity as

$$\varepsilon^{n_1}, \varepsilon^{n_2}, \ldots, \varepsilon^{n_k},$$

where $1 = n_1, n_2, \ldots, n_k$ are those integers less than m which are relatively prime to m, and $k = \text{degree } \Phi_m$. Now, if n_i and m are relatively prime, $\mathbb{Z} = (n_i, m)$ and so there exist integers a and b such that $an_i + bm = 1$. Thus in the quotient ring $\mathbb{Z}_m$, $\bar{a}\bar{n}_i = \bar{1}$, and n_i is a unit. Conversely if $\bar{n}$ is a unit in $\mathbb{Z}_m$ then n and m are relatively prime. Thus $\{\bar{n}_1, \ldots, \bar{n}_k\}$ is the *multiplicative* group U_m of units in the ring $\mathbb{Z}_m$.

Now suppose that σ is in the Galois group $\Gamma_K(\Phi_m)$. As $L = K(\varepsilon)$, σ is determined by its action on ε. As $\sigma(\varepsilon)$ is also a primitive mth root of unity, $\sigma(\varepsilon) = \varepsilon^{n_{j(\sigma)}}$ for some $1 \leqslant j(\sigma) \leqslant k$. If τ is another element of $\Gamma_K(\Phi_m)$,

$$\tau\sigma(\varepsilon) = \tau(\varepsilon^{n_{j(\sigma)}}) = (\tau(\varepsilon))^{n_{j(\sigma)}} = \varepsilon^{n_{j(\tau)}n_{j(\sigma)}} = \sigma\tau(\varepsilon).$$

Thus $\Gamma_K(\Phi_m)$ is abelian. Also

$$\bar{n}_{j(\tau\sigma)} = \bar{n}_{j(\tau)}\bar{n}_{j(\sigma)}$$

and so the mapping $\sigma \rightarrow \bar{n}_{j(\sigma)}$ is a homomorphism of $\Gamma_K(\Phi_m)$ into U_m. This is one–one, since $\sigma(\varepsilon) = \varepsilon$ if and only if σ is the identity in $\Gamma_K(\Phi_m)$. Further, $|\Gamma_K(\Phi_m)| = k$ if and only if there are k images $\varepsilon^{n_{j(\sigma)}}$; thus the homomorphism is onto if and only if $\Gamma_K(\Phi_m)$ acts transitively on the roots of Φ_m, and this happens if and only if Φ_m is irreducible over K. Summing up:

Theorem 15.4 *If Φ_m is the mth cyclotomic polynomial over K, $\Gamma_K(\Phi_m)$ is an abelian group which is isomorphic to a subgroup of U_m, the multiplicative group of units of the ring $\mathbb{Z}_m$. Φ_m is irreducible over K if and only if $\Gamma_K(\Phi_m)$ is isomorphic to U_m.*

As an example,

$$U_{12} = \{\bar{1}, \bar{5}, \bar{7}, \overline{11}\}$$

and $\bar{1}^2 = \bar{5}^2 = \overline{11}^2 = \bar{1}$, so that $U_{12} \cong \mathbb{Z}_2 \times \mathbb{Z}_2$.

If p is a prime, U_p is cyclic, by Theorem 12.3. We therefore have the following corollary:

Corollary *If p is a prime then either Φ_p splits over K or $\Gamma_K(\Phi_p)$ is cyclic.*

Exercises

15.5 Find the Galois groups of x^4+1 and x^5+1 over $\mathbb{Q}$.

15.6 Suppose that p is a prime which does not divide m, and let ε be a primitive mth root of unity over $\mathbb{Z}_p$. Show that $[\mathbb{Z}_p(\varepsilon):\mathbb{Z}_p]=k$, where k is the order of $\bar{p}$ in the multiplicative group U_m of units in $\mathbb{Z}_m$. Show that Φ_m is irreducible over $\mathbb{Z}_p$ if and only if U_m is a cyclic group generated by $\bar{p}$. When is Φ_4 irreducible over $\mathbb{Z}_p$? When is Φ_8 irreducible over $\mathbb{Z}_p$?

15.7 Suppose that $m=q^t$, where q is an odd prime.
  (i) Show that $|U_m|=(q-1)q^{t-1}$.
  (ii) Use the fact that U_q is cyclic of order $q-1$ to show that there is an element of order $q-1$ in U_m.
  (iii) Show that if q does not divide a then
  $$(1+aq^u)^q=1+bq^{u+1},$$
  where q does not divide b.
  (iv) Show that $\overline{1+q}$ has order q^{t-1} in U_m.
  (v) Combine (ii) and (iv) to show that U_m is cyclic.

15.8 Suppose that $m=m_1 \ldots m_r$, where $m_1,\ldots,m_r$ are distinct prime powers. Show that
  $$U_m \cong U_{m_1} \times \cdots \times U_{m_r}.$$

15.9 Show that U_m is cyclic if and only if $m=q^t$ or $2q^t$ (where q is an odd prime) or 4.

15.10 Is Φ_{18} irreducible over (a) $\mathbb{Z}_{23}$, (b) $\mathbb{Z}_{43}$, (c) $\mathbb{Z}_{73}$?

16

Cyclic extensions

If we are going to study extensions by radicals, it is clearly useful to be able to answer the question: when is a Galois extension $L:K$ a splitting field extension for a polynomial of the form $x^n - \theta$?

16.1   A necessary condition

We begin by considering a polynomial of the form $x^n - \theta$ in $K[x]$. In order to avoid problems of separability, let us suppose that char K does not divide n. Let $L:K$ be a splitting field extension for $f = x^n - \theta$ over K. Then by Theorem 10.5, f has n distinct roots, $\alpha_1, \ldots, \alpha_n$ say, in L. Since $(\alpha_i \alpha_j^{-1})^n = \theta\theta^{-1} = 1$, the elements $\alpha_1 \alpha_1^{-1}$, $\alpha_2 \alpha_1^{-1}, \ldots, \alpha_n \alpha_1^{-1}$ are n distinct roots of unity in L, so that $x^n - 1$ splits over L. Let ω be a primitive nth root of unity. Then

$$x^n - \theta = (x - \alpha_1)(x - \omega\alpha_1) \ldots (x - \omega^{n-1}\alpha_1).$$

This suggests that we should consider the intermediate field $K(\omega)$, which contains all the nth roots of unity.

Theorem 16.1 *Suppose that $x^n - \theta \in K[x]$, and that char K does not divide n. Let $L:K$ be a splitting field extension for $x^n - \theta$ over K. Then L contains a primitive nth root of unity, ω say. The group $\Gamma(L:K(\omega))$ is cyclic, and its order divides n. $x^n - \theta$ is irreducible over $K(\omega)$ if and only if $[L:K(\omega)] = n$.*
Proof. We have seen that L contains a primitive nth root of unity and that $x^n - \theta$ splits over L as

$$(x - \beta)(x - \omega\beta) \ldots (x - \omega^{n-1}\beta).$$

Thus $L = K(\omega, \beta)$ and if $\sigma \in \Gamma(L:K(\omega))$, σ is determined by its action on β. Now if σ is in $\Gamma(L:K(\omega))$

$$\sigma(\beta) = \omega^{j(\sigma)}\beta$$

for some $0 \leqslant j(\sigma) < n$. Since $\omega \in K(\omega)$, if σ and τ are in $\Gamma(L:K(\omega))$,

$$\tau\sigma(\beta) = \tau(\omega^{j(\sigma)}\beta) = \omega^{j(\sigma)}\tau(\beta) = \omega^{j(\sigma)}\omega^{j(\tau)}\beta$$

and so the map $\sigma \to \overline{j(\sigma)}$ is a homomorphism of $\Gamma(L:K(\omega))$ into the *additive* group $(\mathbb{Z}_n, +)$. As $\overline{j(\sigma)} = \overline{0}$ if and only if $\sigma(\beta) = \beta$, and this happens if and only if σ is the identity, the homomorphism is one–one. Thus $\Gamma(L:K(\omega))$ is isomorphic to a subgroup of the cyclic group $(\mathbb{Z}_n, +)$: it is therefore cyclic, and its order divides n.

If $x^n - \theta$ is irreducible over $K(\omega)$, $|\Gamma(L:K(\omega))| \geqslant n$, so that $[L:K(\omega)] = |\Gamma(L:K(\omega))| = n$. If $x^n - \theta$ is not irreducible over $K(\omega)$, let g be an irreducible monic factor in $K(\omega)$, and let γ be a root of g in L. Then

$$x^n - \theta = (x - \gamma)(x - \omega\gamma)\ldots(x - \omega^{n-1}\gamma)$$

so that $x^n - \theta$ splits over $K(\omega, \gamma)$, and $L = K(\omega, \gamma)$. Thus

$$[L:K(\omega)] = [K(\omega, \gamma):K(\omega)] = \text{degree } g < n.$$

Exercises

16.1 Show that $x^6 + 3$ is irreducible over $\mathbb{Q}$, but is not irreducible over $\mathbb{Q}(\omega)$, where ω is a primitive sixth root of unity.

16.2 Show that the Galois group of $x^{15} - 2$ over $\mathbb{Q}$ can be generated by elements ρ, σ and τ satisfying

$$\rho^{15} = \sigma^4 = \tau^2 = 1,$$

$$\sigma^{-1}\rho\sigma = \rho^7,$$

$$\tau^{-1}\rho\tau = \rho^{14},$$

$$\tau^{-1}\sigma\tau = \sigma.$$

16.3 Let $L:\mathbb{Q}$ be a splitting field extension for $x^4 - 5$ over $\mathbb{Q}$. What is its Galois group? List the fields intermediate between L and $\mathbb{Q}$, and determine which of them are normal over $\mathbb{Q}$.

16.2   Abel's theorem

In the case where n is a prime, we can say more about the irreducibility of $x^n - \theta$.

Theorem 16.2 (Abel's theorem) *Suppose that q is a prime, that $x^q - \theta \in K[x]$ and that char $K \neq q$. Then either $x^q - \theta$ is irreducible over K or $x^q - \theta$ has a root in K. In the latter case $x^q - \theta$ splits over K if and only if K contains a primitive qth root of unity.*

Proof. Suppose that $x^q - \theta$ is not irreducible over K. Let $L:K$ be a splitting field extension for $x^q - \theta$, let g be an irreducible monic divisor of $x^q - \theta$ in $K[x]$ and let γ be a root of g in L. Then, in L,

$$g = (x - \gamma)(x - \omega^{n_2}\gamma)\ldots(x - \omega^{n_d}\gamma)$$

where ω is a primitive qth root of unity in L, $1 \leqslant n_2 < n_3 < \cdots < n_d < q$ and $d = \text{degree } g$. Thus if

$$g = x^d - g_{d-1}x^{d-1} + \cdots + (-1)^d g_0,$$

$g_0 = \omega^k \gamma^d$ for some k. Raising this to the qth power, we see that $g_0^q = \gamma^{dq} = \theta^d$. Now d and q are coprime, and so there exist integers a and b such that

$$ad + bq = 1.$$

Thus

$$\theta = \theta^{ad}\theta^{bq} = (g_0^a \theta^b)^q$$

and so $x^q - \theta$ has a root $g_0^a \theta^b$ in K.

If $x^q - \theta$ is not irreducible over K, $[L : K(\omega)]$ divides q and is less than q, by Theorem 16.1, and so $L = K(\omega)$. Thus $K(\omega) : K$ is a splitting field extension for $x^q - \theta$: the last statement of the theorem follows immediately from this.

Exercises

16.4 Suppose that q is a prime, that $\text{char } K \neq q$ and that $x^q - \theta$ is irreducible in $K[x]$. Let ω be a primitive qth root of unity, and let $[K(\omega) : K] = j$. Show that the Galois group of $x^q - \theta$ can be generated by elements σ and τ satisfying

$$\sigma^q = \tau^j = 1, \quad \sigma^k \tau = \tau\sigma,$$

where $\bar{k}$ is a generator of the multiplicative group $\mathbb{Z}_q^*$.

16.5 Suppose that q is a prime, that $\text{char } K = q$ and that $\theta \in K$. Describe the splitting field for $x^q - \theta$ over K.

16.3 A sufficient condition

In order to prove a converse to Theorem 16.1 we need a fundamental result, of interest in its own right.

Suppose that G is a group and that K is a field. A (*K-valued*) *character* on G is a homomorphism of G into the multiplicative group K^* of non-zero elements of K. We can think of a character as a K-valued function on G; recall that the set of all K-valued functions on G is a vector space over K.

Theorem 16.3 *Suppose that G is a group, that K is a field and that S is a set of K-valued characters on G. Then S is linearly independent over K.*

Proof. If not, there is a minimal non-empty subset $\{\gamma_1, \ldots, \gamma_n\}$ of distinct elements of S which is linearly dependent over K. That is, there exist non-zero $\lambda_1, \ldots, \lambda_n$ in K such that

$$\lambda_1 \gamma_1(g) + \cdots + \lambda_n \gamma_n(g) = 0 \qquad (*)$$

for all g in G. Each γ_i is non-zero, since it sends the identity of G to 1, and so $n \geqslant 2$. As $\gamma_1 \neq \gamma_n$, there exists h in G such that $\gamma_1(h) \neq \gamma_n(h)$. Now

$$\lambda_1 \gamma_1(hg) + \cdots + \lambda_n \gamma_n(hg) = 0$$

for all g in G. Using the fact that the γ_i are characters, we have that

$$\lambda_1 \gamma_1(h)\gamma_1(g) + \cdots + \lambda_n \gamma_n(h)\gamma_n(g) = 0$$

for all g in G. Now multiply $(*)$ by $\gamma_n(h)$ and subtract:

$$\lambda_1(\gamma_1(h) - \gamma_n(h))\gamma_1(g) + \cdots + \lambda_{n-1}(\gamma_{n-1}(h) - \gamma_n(h))\gamma_{n-1}(g) = 0$$

for all g in G. As $\gamma_1(h) - \gamma_n(h) \neq 0$, this means that $\{\gamma_1, \ldots, \gamma_{n-1}\}$ is linearly dependent over K, contradicting the minimality of $\{\gamma_1, \ldots, \gamma_n\}$.

If τ is an automorphism of a field K, then the restriction of τ to K^* is a K-valued character on K^*. Spelling the theorem out in detail in this case, we have the following corollary:

Corollary *Suppose that $\tau_1, \ldots, \tau_n$ are distinct automorphisms of a field K and that $k_1, \ldots, k_n$ are non-zero elements of K. Then there exists k in K such that*

$$k_1 \tau_1(k) + \cdots + k_n \tau_n(k) \neq 0.$$

We now turn to the converse of Theorem 16.1. We say that an extension $L:K$ is *cyclic* if it is a Galois extension and $\Gamma(L:K)$ is a cyclic group.

Theorem 16.4 *Suppose that $L:K$ is a cyclic extension of degree n, that char K does not divide n and that K contains a primitive nth root of unity, ω say. Then there exists θ in K such that $x^n - \theta$ is irreducible over K and $L:K$ is a splitting field extension for $x^n - \theta$. If β is a root of $x^n - \theta$ in L, then $L = K(\beta)$.*

Proof. Let σ be a generator for the cyclic group $\Gamma(L:K)$. Since the identity, $\sigma, \sigma^2, \ldots, \sigma^{n-1}$ are distinct automorphisms of L, by the corollary to Theorem 16.3 there exists α in L such that

$$\beta = \alpha + \omega\sigma(\alpha) + \cdots + \omega^{n-1}\sigma^{n-1}(\alpha) \neq 0.$$

Observe that $\sigma(\beta) = \omega^{-1}\beta$: this means first that $\beta \notin K$ and secondly that $\sigma(\beta^n) = (\sigma(\beta))^n = \beta^n$, so that $\theta = \beta^n \in K$.

As

$$x^n - \theta = (x - \beta)(x - \omega\beta) \ldots (x - \omega^{n-1}\beta),$$

$K(\beta):K$ is a splitting field extension for $x^n - \theta$ over K. Since the identity,

$\sigma, \ldots, \sigma^{n-1}$ are distinct automorphisms of $K(\beta)$ which fix K,

$$[K(\beta):K] = |\Gamma(K(\beta):K)| \geqslant n$$

and so $L = K(\beta)$. The irreducibility of $x^n - \theta$ over K now follows from Theorem 16.1.

Exercises

16.6 Suppose that $[L:K]$ is a prime p, that $p \neq \operatorname{char} K$ and that L is algebraically closed. Suppose (if possible) that $p > 2$.

   (i) Show that the cyclotomic polynomials Φ_p and Φ_{p^2} split over K.

   (ii) Show that there exists θ in K such that $x^p - \theta$ is irreducible over K and $L:K$ is the splitting field extension for $x^p - \theta$.

   (iii) Show that $f = x^{p^2} - \theta$ has no roots in K, and must be of the form $f = f_1 \ldots f_p$, where each f_j is an irreducible polynomial in $K[x]$ of degree p.

   (iv) Show that if $\alpha_1, \ldots, \alpha_p$ are roots of f_1 then $\alpha_1 \ldots \alpha_p = \omega\beta$, where ω is a p^2th root of unity and $\beta^p = \theta$. Explain why this gives a contradiction.

16.7 Suppose that $[L:K] = 4$, that $\operatorname{char} K \neq 2$ and that L is algebraically closed. Show that there exists an intermediate field M such that $[L:M] = 2$ and such that Φ_4 splits over M. Show that this leads to a contradiction.

16.8 Suppose that $\operatorname{char} K = 0$, that $1 < [L:K] < \infty$ and that L is algebraically closed. Show that $[L:K] = 2$ and that $L:K$ is a splitting field extension for $x^2 + 1$. (You will probably need the fact that if p is a prime which divides the order of a group G then G has a subgroup of order p.)

The next three exercises are concerned with cyclic extensions of degree p, in the case where $\operatorname{char} K = p$.

16.9 Suppose that $\operatorname{char} K = p$, that $f = x^p - x - \alpha \in K[x]$ and that $L:K$ is a splitting field extension for f. Show that if β is a root of f then the roots of f are $\beta, \beta + 1, \ldots, \beta + p - 1$. Show that either f splits over K or f is irreducible over K and $L:K$ is cyclic of degree p.

16.10 Suppose that $L:K$ is a Galois extension with Galois group G. If $x \in L$, let

$$\operatorname{tr}(x) = \sum_{\sigma \in G} \sigma(x).$$

Show that tr is a K-linear mapping of L onto K. The mapping tr is the *trace*. What is the effect of tr on K if char $K \big| \big| G \big|$?

16.11 Suppose that char $K = p$, that $L:K$ is a cyclic extension of degree p and that τ generates $\Gamma(L:K)$. Let z be an element of L with $\mathrm{tr}(z) = 1$, and let

$$y = (p-1)z + (p-2)\tau(z) + \cdots + 2\tau^{p-3}(z) + \tau^{p-2}(z).$$

Show that $\tau(y) - y = 1$, and that $\alpha = y^p - y \in K$. Show that $f = x^p - x - \alpha$ is irreducible over K, that $L:K$ is a splitting field extension for f and that $L = K(y)$.

16.12 Suppose that $L:K$ is a Galois extension of degree n with Galois group G. If $x \in L$, let

$$\mathrm{tr}(x) = \sum_{\sigma \in G} \sigma(x), \quad N(x) = \prod_{\sigma \in G} \sigma(x).$$

The mapping N is the *norm*. Suppose that $\alpha \in L$ has minimal polynomial

$$x^r - a_1 x^{r-1} + \cdots + (-1)^r a_r$$

Show that $\mathrm{tr}(\alpha) = (n/r)a_1$ and $N(\alpha) = a_r^{n/r}$.

16.13 (Hilbert's theorem 90) Suppose that $L:K$ is a Galois extension of degree n with cyclic Galois group generated by τ, say.
  (i) Suppose that $\alpha = \beta/\tau(\beta)$. Show that $N(\alpha) = 1$.
  (ii) Suppose that $N(\alpha) = 1$. Let $c_0 = \alpha$, $c_1 = \alpha\tau(c_0)$, $c_2 = \alpha\tau(c_1), \ldots,$ $c_{n-1} = \alpha\tau(c_{n-2})$. Show that there exists γ in L such that

$$\beta = c_0\gamma + c_1\tau(\gamma) + \cdots + c_{n-1}\tau^{n-1}(\gamma) \neq 0.$$

Show that $\alpha = \beta/\tau(\beta)$.

16.14 Suppose that $L:K$ is a Galois extension of degree n with cyclic Galois group generated by τ, say, and that $\alpha \in L$. Show that $\mathrm{tr}(\alpha) = 0$ if and only if there exists β in L such that $\alpha = \beta - \tau\beta$.

16.4 Kummer extensions

It is not difficult to extend Theorems 16.1 and 16.4 to Galois extensions with abelian Galois groups.

Theorem 16.5 *Suppose that $L:K$ is a Galois extension, that $\Gamma(L:K)$ is an abelian group of exponent d and that $x^d - 1$ has d distinct roots in K. Then there exist $\theta_1, \ldots, \theta_r$ in K such that $L:K$ is a splitting field extension for*

$$(x^d - \theta_1) \ldots (x^d - \theta_r).$$

Proof. The proof is by induction on $[L:K]$. Suppose that $[L:K]=n$ and that the result holds for all extensions of smaller degree.

By Theorem 12.2 we can write

$$\Gamma(L:K)=F \times H$$

where F is cyclic of order d and H is an abelian group whose exponent e divides d. Let M be the fixed field of F; then $M:K$ is a Galois extension, and $\Gamma(M:K) \cong H$. As x^e-1 has e distinct roots in K and $[M:K]<n$, by the inductive hypothesis there exist $\psi_1,\ldots,\psi_{r-1}$ in K such that $M:K$ is a splitting field extension for

$$(x^e-\psi_1)\ldots(x^e-\psi_{r-1}).$$

Let β_j be a root of $x^e-\psi_j$ in M, let $\theta_j=\psi_j^{d/e}$ and let ω be a primitive dth root of unity in K. Then

$$x^d-\theta_j=\prod_{i=0}^{d-1}(x-\omega^i\beta_j)$$

so that $M:K$ is a splitting field extension for

$$(x^d-\theta_1)\ldots(x^d-\theta_{r-1}).$$

We now argue as in Theorem 16.4. Let σ be a generator for F. As $\Gamma(L:K)=\{\sigma^j\tau:0 \leqslant j<d, \tau \in H\}$, there exists α in L such that

$$\beta_r=\sum_{\tau \in H}\tau(\alpha)+\omega\sigma\sum_{\tau \in H}\tau(\alpha)+\cdots+\omega^{d-1}\sigma^{d-1}\sum_{\tau \in H}\tau(\alpha) \neq 0.$$

As before, $\sigma(\beta_r)=\omega^{-1}\beta_r$, so that $\sigma(\beta_r^d)=(\sigma(\beta_r))^d=\beta_r^d$ and $\theta_r=\beta_r^d \in M$. But also $\tau(\beta_r)=\beta_r$ for $\tau \in H$, so that $\tau(\theta_r)=\theta_r$ for $\tau \in H$, and so $\theta_r \in K$. As in Theorem 16.4, $L:M$ is a splitting field extension for $x^d-\theta_r$, and so $L:K$ is a splitting field extension for $(x^d-\theta_1)\ldots(x^d-\theta_r)$.

As extension $L:K$ is called a *Kummer extension of exponent d* if it is a splitting field extension of a polynomial of the form

$$(x^d-\theta_1)\ldots(x^d-\theta_r)$$

(where $\theta_1,\ldots,\theta_r$ are in K) and if x^d-1 has d distinct roots in K.

Let us now prove a converse to Theorem 16.5.

Theorem 16.6 *Suppose that $L:K$ is a Kummer extension of exponent d. Then $\Gamma(L:K)$ is abelian, and its exponent divides d.*

Proof. Suppose that $L:K$ is a splitting field extension of

$$f=(x^d-\theta_1)\ldots(x^d-\theta_r).$$

By Theorem 11.6, we need only consider the action of $\Gamma(L:K)$ on the roots of f. Let ω be a primitive dth root of unity in K. Then if $\sigma \in \Gamma(L:K)$ and β_j is a root of $x^d-\theta_j$ in L, $\sigma(\beta_j)=\omega^{n_{\sigma,j}}\beta_j$ for some $n_{\sigma,j}$, so that $\sigma^d(\beta_j)=\beta_j$, and $\sigma^d=e$.

This implies that the exponent of $\Gamma(L:K)$ divides d. If τ is another element of $\Gamma(L:K)$,

$$\tau\sigma(\beta_j) = \tau(\omega^{n_\sigma j}\beta_j) = \omega^{n_\sigma j}\omega^{n_\tau j}\beta_j$$
$$= \sigma(\omega^{n_\tau j}\beta_j) = \sigma\tau(\beta_j),$$

so that $\Gamma(L:K)$ is abelian.

Exercises

16.15 Suppose that K is a field which contains a primitive nth root of unity and that $x^n - a$ and $x^n - b$ are irreducible over K. Show that if $b = a^r c^n$ for some r which is prime to n and some c in K, then $x^n - a$ and $x^n - b$ have the same splitting field extension over K.

16.16 Suppose K is a field which contains a primitive nth root of unity, and that $x^n - a$ and $x^n - b$ are irreducible polynomials over K with the same splitting field extension $L:K$. Let α be a root of $x^n - a$ in L, β a root of $x^n - b$. By considering the action of $\Gamma(L:K)$ on α and β, show that there exists r, prime to n, such that $\beta\alpha^{-r} \in K$. Show that $b = a^r c^n$ for some c in K.

17

Solution by radicals

The results of the two preceding chapters, together with the fundamental theorem of Galois theory, suggest that, provided that we can construct enough roots of unity, a separable polynomial is solvable by radicals if and only if its Galois group can be built up in some way from abelian groups. We shall see that this is indeed so: but first we must develop some group theory.

17.1 Soluble groups: examples

A group G is said to be *soluble* if there is a finite series of subgroups

$$\{e\} = G_n \subseteq G_{n-1} \subseteq \cdots \subseteq G_1 \subseteq G_0 = G$$

such that

(i) $G_i \lhd G_{i-1}$ for $1 \leqslant i \leqslant n$, and
(ii) G_{i-1}/G_i is cyclic, for $1 \leqslant i \leqslant n$.

Let us consider some examples. The alternating group A_3 is a cyclic normal subgroup of Σ_3, and Σ_3/A_3 is cyclic of order 2, so that A_3 and Σ_3 are soluble. In Σ_4 let $G_4 = \{e\}$, $G_3 = \{e, (12)(34)\}$, $G_2 = N$, the Viergruppe, $G_1 = A_4$ and $G_0 = \Sigma_4$. Then

$$G_4 \lhd G_3 \lhd G_2 \lhd G_1 \lhd G_0$$

and $G_3/G_4, G_2/G_3$ and G_0/G_1 are all cyclic of order 2, while G_1/G_2 is cyclic of order 3. Thus A_4 and Σ_4 are soluble.

If G is a finitely generated abelian group, generated by $g_1, \ldots, g_n$ say, let $G_n = \{e\}$ and $G_j = \langle g_1, \ldots, g_{n-j} \rangle$ for $0 \leqslant j < n$. Then trivially $G_j \lhd G_{j-1}$ for $1 \leqslant j \leqslant n$, and each G_{j-1}/G_j is cyclic, since G_{j-1}/G_j is generated by the coset $G_j + g_{n-j+1}$. Thus every finitely generated abelian group (and in particular every finite abelian group) is soluble.

We have seen that A_3 and A_4 are soluble. We shall show that the alternating groups A_n, for $n \geqslant 5$, are all *simple*: that is, they have no non-trivial normal subgroups. Since these groups are certainly not cyclic, it follows that they are not soluble.

Let us show that A_n is simple for $n \geqslant 5$. It will be assumed that you are familiar with the representations of permutations as products of disjoint cycles. A_n is generated by the set S of products of two transpositions. We can write $S = \{e\} \cup B \cup T$, where B is the set of products of two disjoint transpositions and T is the set of 3-cycles. Now if $n \geqslant 5$, any element of T can be written as a product of two elements of B: for example, $(123) = ((45)(12))((23)(45))$. It is therefore sufficient to show that if N is a normal subgroup of A_n other than $\{e\}$, then $N \supseteq B$.

We prove this by induction on n, for $n \geqslant 4$. The result is true for $n = 4$, since the normal subgroups of A_4 are $\{e\}$, the Viergruppe and A_4. Suppose that $n > 4$, that the result is true for $n - 1$, and that N is a normal subgroup of A_n other than $\{e\}$. For each $1 \leqslant j \leqslant n$, let $F_j = \{\sigma \in A_n : \sigma(j) = j\}$. Each F_j is a subgroup of A_n, isomorphic to A_{n-1}. Suppose that τ is an element of N other than the identity. If τ is a product of disjoint transpositions – $\tau = (a_1 a_2)(a_3 a_4) \ldots$ say – then

$$(a_1 a_3)(a_2 a_4) = (a_1 a_2 a_3)\tau(a_1 a_3 a_2)\tau \in N.$$

Otherwise τ contains a cycle of length greater than 2, so that we can write $\tau = (a_1 a_2 a_3 \ldots a_k)\sigma$. Let b be different from a_1, a_2 and a_3. Then either $\tau \in F_b$ or, setting $\tau' = (a_2 \tau(b))(a_1 b)\tau(a_1 b)(a_2 \tau(b))$, $\tau' \in N$, $\tau'(b) = \tau(b)$, while $\tau'(\tau(b)) = a_3 \neq \tau(\tau(b))$ so that $\tau^{-1}\tau'$ is an element of $N \cap F_b$ other than the identity.

In either case, we conclude that there exists j such that $N \cap F_j \neq \{e\}$. But $N \cap F_j$ is a normal subgroup of F_j and so, by the inductive hypothesis, N contains all elements of B which fix j. But if $(ij)(kl)$ is an element of B which moves j, and if m is different from i, j, k and l then

$$(ij)(kl) = ((mj)(kl))((im)(kl))((mj)(kl)) \in N.$$

Thus $N \supseteq B$ and the proof is complete.

17.2   Soluble groups: basic theory

Let us now establish some results concerning soluble groups. First we need a general result.

Theorem 17.1 *Suppose that G is a group, that $H \lhd G$, and that A is a subgroup of G.*

  (i) *$H \cap A \lhd A$, and $A/(H \cap A) \cong HA/A$.*
  (ii) *If $H \subseteq A$ and $A \lhd G$, then $H \lhd A$, $A/H \lhd G/H$ and $(G/H)/(A/H) \cong G/A$.*

Proof. (i) If $h \in H \cap A$ and $a \in A$, $a^{-1}ha \in H \cap A$, so that $H \cap A \lhd A$. Let q be the quotient mapping $G \to G/H$, and let $j: A \to G$ be the inclusion mapping.

The qj is a homomorphism of A onto HA/H with kernel $H \cap A$, so that the result follows from the first isomorphism theorem.

(ii) Certainly $H \lhd A$, by (i). If C is a right coset of H in G, let $\theta(C) = AC$. AC is a right coset of A in G and $\theta : G/H \to G/A$ is a homomorphism onto G/A. The kernel is $\{C : C \subseteq A\} = A/H$, so that the result again follows from the first isomorphism theorem.

Theorem 17.2 (i) *If G is a soluble group and A is a subgroup of G then A is soluble.*

(ii) *Suppose that G is a group and $H \lhd G$. Then G is soluble if and only if H and G/H are soluble.*

Proof. (i) Let $\{e\} = G_n \lhd G_{n-1} \lhd \cdots \lhd G_0 = G$, such that G_{i-1}/G_i is cyclic for $1 \leqslant i \leqslant n$. Let $A_i = A \cap G_i$. Then $A_i = A_{i-1} \cap G_i \lhd A_{i-1}$ and

$$A_{i-1}/A_i \cong G_i A_{i-1}/G_i$$

by (i) of Theorem 17.1. But $G_i A_{i-1}/G_i$ is a subgroup of the cyclic group G_{i-1}/G_i, so that either $A_{i-1} = A_i$ or A_{i-1}/A_i is cyclic.

(ii) First suppose that G is soluble, and that $\{e\} = G_n \lhd \cdots \lhd G_0 = G$, with G_{i-1}/G_i cyclic. H is soluble, by (i). Let $H_i = HG_i/H$. Now if $Hg_i \in H_i$ and $Hg_{i-1} \in H_{i-1}$,

$$(Hg_{i-1})^{-1} Hg_i Hg_{i-1} = g_{i-1}^{-1} Hg_i Hg_{i-1}$$
$$= Hg_{i-1}^{-1} g_i g_{i-1} \in H_i$$

so that $H_i \lhd H_{i-1}$.

Let $q : G \to G/H$ be the quotient mapping and let $q_i : H_{i-1} \to H_{i-1}/H_i$ denote the quotient mapping. Then $q(G_{i-1}) = H_{i-1}$, and $q_i q$ maps G_{i-1} onto H_{i-1}/H_i. An element g of G_{i-1} is in the kernel of $q_i q$ if and only if $Hg \in HG_i$: that is, if and only if $g \in HG_i$. The restriction of $q_i q$ to G_{i-1} therefore has kernel $HG_i \cap G_{i-1}$ so that by the first isomorphism theorem,

$$H_{i-1}/H_i \cong G_{i-1}/(HG_i \cap G_{i-1}).$$

But $G_i \lhd G_{i-1}$, and $G_i \subseteq HG_i \cap G_{i-1}$, so that by Theorem 17.1(ii),

$$G_{i-1}/(HG_i \cap G_{i-1}) \cong (G_{i-1}/G_i)/((HG_i \cap G_{i-1})/G_i).$$

The right-hand side is a quotient of a cyclic group, so that either $H_{i-1} = H_i$ or H_{i-1}/H_i is cyclic. Thus G/H is soluble.

Now suppose that $H \lhd G$ and that H and G/H are soluble. There exist

$$\{e\} = H_n \lhd \cdots \lhd H_0 = H, \text{ with } H_{i-1}/H_i \text{ cyclic}$$

and

$$\{H\} = K_m \lhd K_{m-1} \lhd \cdots \lhd K_0 = G/H, \text{ with } K_{j-1}/K_j \text{ cyclic}.$$

Let $q : G \to G/H$ be the quotient mapping, and let $G_j = q^{-1}(K_j)$ for $0 \leqslant j \leqslant m$.

If $g_j \in G_j$ and $g_{j-1} \in G_{j-1}$

$$q(g_{j-1}^{-1} g_j g_{j-1}) = (q(g_{j-1}))^{-1} q(g_j) q(g_{j-1}) \in K_j,$$

since $K_j \lhd K_{j-1}$, and so $g_{j-1}^{-1} g_j g_{j-1} \in G_j$. Thus $G_j \lhd G_{j-1}$. By Theorem 17.1(ii)

$$G_{j-1}/G_j \cong (G_{j-1}/H)/(G_j/H) = K_{j-1}/K_j,$$

which is cyclic. Thus the series

$$\{e\} = H_n \lhd \cdots \lhd H_0 = H = G_m \lhd G_{m-1} \lhd \cdots \lhd G_0 = G$$

shows that G is soluble.

Corollary *Σ_n is not soluble, for $n \geqslant 5$.*

It has a subgroup (A_5) which is not soluble.

Exercises

17.1 A group G is *nilpotent* if there is a finite series of subgroups

$$\{e\} = G_n \subseteq G_{n-1} \subseteq \cdots \subseteq G_0 = G$$

such that each G_i is normal in G and G_i/G_{i+1} is abelian for $0 \leqslant i < n$. Show that a finite nilpotent group is soluble and give an example of a finite soluble group which is not nilpotent.

17.2 Suppose that G is a group of order p^n (where p is a prime). Show that the centre $Z = \{z : gz = zg$ for all g in $G\}$ has at least p elements, and show that G is nilpotent.

17.3   Polynomials with soluble Galois groups

In this section we shall show that, if f is separable and has a soluble Galois group, then, provided that we can construct enough roots of unity, f is solvable by radicals.

Theorem 17.3 *Suppose that f is a separable polynomial in $K[x]$ whose Galois group $\Gamma_K(f)$ is soluble, and suppose that char K does not divide $|\Gamma_K(f)|$. Then f is solvable by radicals.*

Proof. The proof is largely a matter of putting together results that we have already established. Let $d = |\Gamma_K(f)|$. If K does not contain a primitive dth root of unity, we can adjoin one, ε say. Let $L = K(\varepsilon)$. $L:K$ is an extension by radicals. Since char K does not divide d, L contains d distinct dth roots of unity. Now let $N:L$ be a splitting field extension for f over K. $N:L$ is a Galois extension, and by the theorem on natural irrationalities (Theorem 11.9) $\Gamma(N:L) = \Gamma_L(f)$ is isomorphic to a subgroup of $\Gamma_K(f)$. Thus $\Gamma_L(f)$ is soluble, by Theorem 17.2(i). This means that there exist subgroups

$$\{e\} = G_r \lhd G_{r-1} \lhd \cdots \lhd G_0 = \Gamma_L(f)$$

such that G_{j-1}/G_j is cyclic, for $1 \leqslant j \leqslant r$.

We now exploit the fundamental theorem of Galois theory. Let L_j be the fixed field of G_j. Then

$$N = L_r : L_{r-1} : \ldots : L_0 = L.$$

Also $\Gamma(N : L_{j-1}) = G_{j-1}$, and $G_j \lhd G_{j-1}$, so that, by the fundamental theorem,

$$\Gamma(L_j : L_{j-1}) \cong G_{j-1}/G_j, \text{ for } 1 \leqslant j \leqslant r.$$

Thus $L_j : L_{j-1}$ is a cyclic extension. Also $[L_j : L_{j-1}] = |G_{j-1}/G_j|$, so that $[L_j : L_{j-1}]$ divides d: thus char K does not divide $[L_j : L_{j-1}]$, and L_{j-1} contains a primitive $[L_j : L_{j-1}]$th root of unity. By Theorem 16.4, there exists an element β_j in L_j such that $L_j = L_{j-1}(\beta_j)$ and such that β_j is a radical over L_{j-1}. Thus $N : L$ is an extension by radicals, and so $N : K$ is also an extension by radicals. Since f splits over N, f is solvable by radicals.

Notice that if $f \in K[x]$ and if either char $K = 0$ or char $K >$ degree f, then f must be separable, by Theorem 10.6, and char K cannot divide $|\Gamma_K(f)|$, since $|\Gamma_K(f)|$ divides (degree f)!.

17.4   Polynomials which are solvable by radicals

We now turn to results in the opposite direction. Here, the main problem is one of normality. Suppose that

$$L = L_r : L_{r-1} : \ldots : L_0 = K$$

is an extension by radicals. Even if K contains sufficiently many roots of unity, so that each of the extensions $L_j : L_{j-1}$ is normal, it does not follow that $L : K$ is a normal extension. We get round this difficulty by a symmetrization argument.

Theorem 17.4 *Suppose that $L : K$ is a Galois extension, that $M = L(\beta)$, where β is a root of $x^n - \theta$ (with θ in L) and that char K does not divide n. Then there exists an extension by radicals $N : M$ such that $N : K$ is a Galois extension.*
Proof. Since char K does not divide n we can if necessary adjoin a primitive nth root of unity, ε say, to M. Then in $M(\varepsilon)[x]$

$$x^n - \theta = (x - \beta)(x - \varepsilon\beta) \ldots (x - \varepsilon^{n-1}\beta)$$

so that $M(\varepsilon) : L$ is a splitting field extension for $x^n - \theta$ over L. As $x^n - \theta$ has n distinct roots, $M(\varepsilon) : L$ is a Galois extension. Note also that $M(\varepsilon) : L = L(\beta, \varepsilon) : L$ is an extension by radicals.

Now let $G = \Gamma(L : K)$ and let

$$f = \prod_{\sigma \in G} (x^n - \sigma(\theta)).$$

Let $N:M(\varepsilon)$ be a splitting field for f over $M(\varepsilon)$: we have the following tower of extensions:

$$N:M(\varepsilon):M:L:K.$$

If β_σ is a root of $x^n - \sigma(\theta)$ in N then

$$x^n - \sigma(\theta) = (x - \beta_\sigma)(x - \varepsilon\beta_\sigma)\ldots(x - \varepsilon^{n-1}\beta_\sigma);$$

thus $N:L$ is a splitting field extension for f over L. Also, $x^n - \sigma(\theta)$ has n distinct roots in N, for each σ, so that f is separable over $M(\varepsilon)$. As $M(\varepsilon):L$ and $L:K$ are both separable, this means that $N:K$ is separable, by Corollary 4 to Theorem 10.3.

We now use the symmetry of f; if $\tau \in G$, $\tau(f) = f$ so that, since $L:K$ is a Galois extension, $f \in K[x]$. There exists g in $K[x]$ such that $L:K$ is a splitting field extension for g over K. Thus $N:K$ is a splitting field extension for fg over K, and so $N:K$ is normal.

Finally observe that N is obtained from M by first adjoining ε and then adjoining the roots of $x^n - \sigma(\theta)$, for σ in G, and so $N:M$ is an extension by radicals.

We now apply this to extensions by radicals.

Theorem 17.5 *Suppose that*

$$L = L_r:L_{r-1}:\ldots:L_0 = K$$

is an extension by radicals, with $L_i = L_{i-1}(\beta_i)$, where β_i is a root of $x^{n_i} - \theta_i$ (with $\theta_i \in L_{i-1}$). Then if char K does not divide $n_1 n_2 \ldots n_r$, there exists an extension $M:L$ such that $M:K$ is a Galois extension by radicals.

Proof. We prove this by induction on r. The result is trivially true when $r = 0$. Suppose that the result holds for $r - 1$. Then there exists an extension $M_{r-1}:L_{r-1}$ such that $M_{r-1}:K$ is a Galois extension by radicals.

Let m_r be the minimal polynomial for β_r over L_{r-1}, and let l_r be an irreducible factor of m_r, considered as an element of $M_{r-1}[x]$. By Theorem 7.2, there is a simple algebraic extension $M_{r-1}(\gamma):M_{r-1}$ such that $l_r(\gamma) = 0$. Since this means that $m_r(\gamma) = 0$, it follows from Theorem 7.4 that there is a monomorphism i from $L = L_{r-1}(\beta_r)$ into $M_{r-1}(\gamma)$, fixing L_{r-1}, such that $i(\beta_r) = \gamma$. In other words, identifying L with $i(L)$, we can suppose that L and M_{r-1} are both subfields of $M_{r-1}(\beta_r)$.

We apply Theorem 17.4 to the Galois extension $M_{r-1}:K$ and $M_{r-1}(\beta_r)$, and conclude that there is an extension $M_r:M_{r-1}(\beta_r)$ by radicals such that $M_r:K$ is a Galois extension. We have the following diagram:

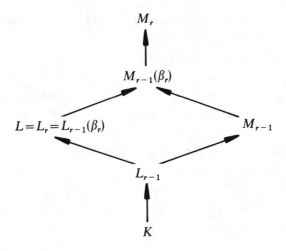

Since each of the extensions $M_r:M_{r-1}(\beta_r)$, $M_{r-1}(\beta_r):M_{r-1}$ and $M_{r-1}:K$ is an extension by radicals, so is $M_r:K$.

Notice that the conditions on char K are also satisfied by the extension $M:K$.

Theorem 17.6 *Suppose that*

$$L=L_r:L_{r-1}:\ldots:L_0=K$$

is an extension by radicals, with $L_i=L_{i-1}(\beta_i)$, where β_i is a root of $x^{n_i}-\theta_i$ (with θ_i in L_{i-1}), and that char K *does not divide $n_1\ldots n_r$. If $f\in K[x]$ splits over L, then the Galois group $\Gamma_K(f)$ is soluble.*

Proof. By Theorem 17.5 and the remark following it, we can assume that $L:K$ is a Galois extension.

For each $1\leqslant i\leqslant r$, $L:L_i$ is a Galois extension (Corollary 3 to Theorem 9.1, and Theorem 10.1): $x^{n_i}-\theta_i$ has a root β_i in L, and so it splits over L. This means, by Theorem 16.1, that L contains a primitive n_ith root of unity. Let n be the lowest common multiple of $n_1,\ldots,n_r$: then L contains a primitive nth root of unity, ε say.

Now let $L_i'=L_i(\varepsilon)$, for $0\leqslant i\leqslant r$. Then we have the following tower of extensions:

$$L=L_r':L_{r-1}':\ldots:L_0'=L_0(\varepsilon):L_0=K.$$

We shall show that $G=\Gamma(L:K)$ is soluble. Let $G_i=\Gamma(L:L_i')$, for $0\leqslant i\leqslant r$. As $L_i':L_{i-1}'$ is a splitting field extension for $x^{n_i}-\theta_i$ over L_{i-1}', $L_i':L_{i-1}'$ is cyclic, and so by the fundamental theorem of Galois theory $G_i\lhd G_{i-1}$ and $G_{i-1}/G_i\cong\Gamma(L_i':L_{i-1}')$. Thus $G_0=\Gamma(L:L_0')=\Gamma(L:K(\varepsilon))$ is soluble. Now ε is a primitive nth root of unity, so that $K(\varepsilon):K$ is a splitting field extension for

$x^n - 1$, and $\Gamma(K(\varepsilon):K)$ is abelian (Theorem 15.4), and therefore soluble. By the fundamental theorem of Galois theory,

$$\Gamma(K(\varepsilon):K) \cong \Gamma(L:K)/\Gamma(L:K(\varepsilon)) = G/G_0$$

so that G/G_0 is soluble. Consequently G is soluble, by Theorem 17.2.

Remember that f splits over L. Let $N:K$ be a splitting field extension for f over K, with $N \subseteq L$. The extension $N:K$ is normal; using the fundamental theorem of Galois theory once again,

$$\Gamma_K(f) = \Gamma(N:K) \cong \Gamma(L:K)/\Gamma(L:N) = G/\Gamma(L:N),$$

and so $\Gamma_K(f)$ is soluble, by Theorem 17.2.

As an example, we have seen that the quintic $x^5 - 4x + 2$ is irreducible over $\mathbb{Q}$, and has Galois group Σ_5. Σ_5 is not soluble, and so $x^5 - 4x + 2$ cannot be solved by radicals!

Exercise

17.3 Let $L_0 = \mathbb{Q}$, $L_1 = \mathbb{Q}(3^{1/2})$, $L_2 = \mathbb{Q}((3^{1/2} + 1)^{1/2})$. Show that $L_1:L_0$ and $L_2:L_1$ are both normal extensions but that $L_2:L_0$ is not normal. Find the minimal polynomial of $(3^{1/2} + 1)^{1/2}$ over $\mathbb{Q}$, and find its Galois group.

17.4 Let f be an irreducible cubic in $K[x]$, where K is a subfield of $\mathbb{R}$. Show that f has three real roots if and only if its discriminant is positive.

17.5 Suppose that K is a subfield of $\mathbb{R}$ and that f is an irreducible cubic in $K[x]$ with three real roots. Suppose that $L = K(r)$, where $r \in \mathbb{R}$ and $r^p \in K$ for some prime p. Show that f is irreducible over L.

17.6 Suppose that K is a subfield of $\mathbb{R}$ and that f is an irreducible cubic in $K[x]$ with three real roots. Show that if $L:K$ is an extension by radicals with $L \subseteq \mathbb{R}$ then f is irreducible over L. (It is not possible to solve f only by extracting real roots!)

17.7 Give an example of a polynomial in $\mathbb{Q}[x]$ which is solvable by radicals, but whose splitting field is not an extension by radicals.

18

Transcendental elements and algebraic independence

18.1 Transcendental elements and algebraic independence

In this chapter, we leave the study of algebraic extensions, and consider problems concerning transcendence.

Suppose that $L:K$ is an extension and that $\alpha \in L$. Recall that α is transcendental over K if the evaluation map $E_\alpha : K[x] \to L$ is one–one; that is, α satisfies no non-zero polynomial relation with coefficients in K.

Theorem 18.1 Suppose that $L:K$ is an extension and that $\alpha \in L$ is transcendental over K. Then the evaluation map E_α can be extended uniquely to an isomorphism F_α from the field $K(x)$ of rational expressions in x over K onto the field $K(\alpha)$.

Proof. The proof should be quite obvious: here are the details.

Remember that the field $K(x)$ is obtained by considering an equivalence relation on $K[x] \times (K[x])^*$ (see Section 3.2).

Suppose that $(f, g) \in K[x] \times (K[x])^*$. As α is transcendental over K, $g(\alpha) \neq 0$, and we can define $G_\alpha(f, g) = f(\alpha)(g(\alpha))^{-1}$. If $(f, g) \sim (f', g')$ then $fg' = f'g$ in $K[x]$, so that $f(\alpha)g'(\alpha) = f'(\alpha)g(\alpha)$ and $G_\alpha(f, g) = G_\alpha(f', g')$. Thus G_α is constant on equivalence classes: we can therefore define $F_\alpha(f/g) = G_\alpha(f, g)$. It is straightforward to verify that F_α is a ring homomorphism. Since $F_\alpha(x) = E_\alpha(x) = \alpha$, $F_\alpha(K(x)) \supseteq K(\alpha)$. On the other hand, if $f/g \in K(x)$, $F_\alpha(f/g) = f(\alpha)(g(\alpha))^{-1} \in K(\alpha)$, and so $F_\alpha(K(x)) = K(\alpha)$. Finally if F'_α is another monomorphism which extends E_α, the set

$$\{r \in K(x) : F_\alpha(r) = F'_\alpha(r)\}$$

is a subfield of $K(x)$ which contains $K[x]$; it must therefore be the whole of $K(x)$, and so F_α is unique.

We now generalize the idea of a transcendental element. Suppose that $L:K$ is an extension and that $A = \{\alpha_1, \ldots, \alpha_n\}$ is a finite subset of L (where

$\alpha_1, \ldots, \alpha_n$ are distinct). Any element f of $K[x_1, \ldots, x_n]$ can be written in the form

$$f = \sum_{j=1}^{m} k_j x_1^{d_{1,j}} \ldots x_n^{d_{n,j}}$$

where $k_j \in K$ for $1 \leqslant j \leqslant m$, and $d_{i,j}$ is a non-negative integer for $1 \leqslant i \leqslant n$, $1 \leqslant j \leqslant m$. We define the *evaluation map* E_A from $K[x_1, \ldots, x_n]$ into L by setting

$$E_A(f) = \sum_{j=1}^{m} k_j \alpha_1^{d_{1,j}} \ldots \alpha_n^{d_{n,j}}.$$

It is easy to see that E_A is a ring homomorphism. We shall frequently write $E_A(f)$ as $f(\alpha_1, \ldots, \alpha_n)$.

We say that A is *algebraically independent* over K if E_A is one–one: that is, there is no polynomial relation, with coefficients in K, between the elements $\alpha_1, \ldots, \alpha_n$. Thus a one-point set $\{\alpha\}$ is algebraically independent over K if and only if α is transcendental over K.

We say that an arbitrary subset S of L is algebraically independent over K if each of its finite subsets is algebraically independent over K.

The proof of the next result is exactly similar to the proof of Theorem 18.1: this time we omit the details.

Theorem 18.2 *Suppose that $L:K$ is an extension and that $A = \{\alpha_1, \ldots, \alpha_n\}$ is algebraically independent over K. Then the evaluation map E_A can be extended uniquely to an isomorphism F_A from the field $K(x_1, \ldots, x_n)$ of rational expressions in $x_1, \ldots, x_n$ onto the field $K(\alpha_1, \ldots, \alpha_n)$.*

The next theorem is again very easy: it gives a useful practical criterion for a finite set to be algebraically independent over K.

Theorem 18.3 *Suppose that $L:K$ is an extension and that $\alpha_1, \ldots, \alpha_n$ are distinct elements of L. Let $K_0 = K$, $K_i = K(\alpha_1, \ldots, \alpha_i)$ for $1 \leqslant i \leqslant n$. Then $A = \{\alpha_1, \ldots, \alpha_n\}$ is algebraically independent over K if and only if α_i is transcendental over K_{i-1}, for $1 \leqslant i \leqslant n$.*

Proof. Suppose that α_i is algebraic over K_{i-1}. Thus

$$f(\alpha_i) = k_0 + k_1 \alpha_i + \cdots + k_r \alpha_i^r = 0$$

for some non-zero f in $K_{i-1}[x]$. We can write each k_j as

$$k_j = p_j(\alpha_1, \ldots, \alpha_{i-1})(q_j(\alpha_1, \ldots, \alpha_{i-1}))^{-1}$$

where the p_j and q_j are in $K[x_1, \ldots, x_{i-1}]$ and the $q_j(\alpha_1, \ldots, \alpha_{i-1})$ are non-zero. We clear the denominators. Let

$$l_j = p_j \left(\prod_{k \neq j} q_k \right), \text{ for } 0 \leqslant j \leqslant r.$$

Then each l_j is in $K[x_1, \ldots, x_{i-1}]$ and

$$g = l_0 + l_1 x_i + \cdots + l_r x_i^r$$

is a non-zero element of $K[x_1, \ldots, x_i]$. As $g(\alpha_1, \ldots, \alpha_i) = 0$, A is not algebraically independent over K.

Conversely suppose that $\{\alpha_1, \ldots, \alpha_n\}$ is not algebraically independent over K. There exists an index j such that $\{\alpha_1, \ldots, \alpha_{j-1}\}$ is algebraically independent over K, while $\{\alpha_1, \ldots, \alpha_j\}$ is not. Thus there exists a non-zero g in $K[x_1, \ldots, x_j]$ such that $g(\alpha_1, \ldots, \alpha_j) = 0$. Grouping terms together, we can write

$$g = k_0 + k_1 x_j + \cdots + k_r x_j^r$$

where $k_i \in K[x_1, \ldots, x_{j-1}]$ for $0 \leqslant i \leqslant r$. Let

$$h = k_0(\alpha_1, \ldots, \alpha_{j-1}) + k_1(\alpha_1, \ldots, \alpha_{j-1})x + \cdots + k_r(\alpha_1, \ldots, \alpha_{j-1})x^r.$$

Then $h \in K_{j-1}[x]$, and h is non-zero, since $\{\alpha_1, \ldots, \alpha_{j-1}\}$ is algebraically independent over K. As $h(\alpha_j) = 0$, α_j is algebraic over K_{j-1}.

Exercises

18.1 Suppose that $L:K$ is an extension, and that $\{\alpha_1, \ldots, \alpha_s\}$ is algebraically independent over K. Show that if $\beta \in K(\alpha_1, \ldots, \alpha_s)$ and $\beta \notin K$ then β is transcendental over K (cf. Exercise 5.5).

18.2 Suppose that $K(\alpha):K$ is a simple extension and that α is transcendental over K. Show that if τ is an automorphism of $K(\alpha)$ which fixes K then there exist a, b, c and d in K with $ad \neq bc$ such that

$\tau(\alpha) = (a\alpha + b)/(c\alpha + d)$.

Conversely show that any such a, b, c and d determine an automorphism of $K(\alpha)$ which fixes K.

18.3 Suppose that $K(\alpha):K$ is a simple extension and that α is transcendental over K. Let σ be the automorphism of $K(\alpha)$ which fixes K and sends α to $1/(1-\alpha)$. Verify that σ^3 is the identity, and determine the fixed field of σ.

18.4 Suppose that $K(\alpha):K$ is a simple extension, that α is transcendental over K, and that char K is an odd prime p. Suppose that $1 < n < p$. Let τ be the automorphism of $K(\alpha)$ which fixes K and sends α to $n\alpha$. Determine the fixed field of τ.

18.2   Transcendence bases

We now introduce an idea which corresponds in many ways to the concept of basis of a vector space. Suppose that $L:K$ is an extension. Let $\mathscr{I}$

denote the collection of all subsets of L which are algebraically independent over K. We order $\mathscr{I}$ by inclusion. An element S of $\mathscr{I}$ which is maximal in this ordering is called a *transcendence basis* for L over K.

The next result characterizes transcendence bases.

Theorem 18.4 *Suppose that $L:K$ is an extension and that S is a subset of L. Then S is a transcendence basis for L over K if and only if S is algebraically independent over K and $L:K(S)$ is algebraic.*

Proof. Suppose that S is a transcendence basis for L over K. Suppose that α is an element of L which is not in S. By the maximality of S, $S \cup \{\alpha\}$ is not algebraically independent over K, so there exist distinct $s_1, \ldots, s_n$ in S and a non-zero f in $K[x_0, \ldots, x_n]$ such that

$$f(\alpha, s_1, \ldots, s_n) = 0.$$

We can write f as

$$k_0 + k_1 x_0 + \cdots + k_j x_0^j,$$

where $k_i \in K[x_1, \ldots, x_n]$, for $0 \leqslant i \leqslant j$, and $k_j \neq 0$. Now $\{s_1, \ldots, s_n\}$ is algebraically independent over K. From this we conclude first that $j \geqslant 1$ and secondly that $k_j(s_1, \ldots, s_n) \neq 0$. Now

$$k_0(s_1, \ldots, s_n) + k_1(s_1, \ldots, s_n)\alpha + \cdots + k_j(s_1, \ldots, s_n)\alpha^j = 0:$$

since $k_i(s_1, \ldots, s_n) \in K(S)$, this means that α is algebraic over $K(S)$, and that $L:K(S)$ is algebraic.

Conversely, suppose that S is algebraically independent over K and that $L:K(S)$ is algebraic. If α is an element of L which is not in S, α is algebraic over $K(S)$, so there exists a non-zero

$$g = k_0 + k_1 + \cdots + k_j x^j$$

in $K(S)[x]$ such that $g(\alpha) = 0$. Each coefficient k_i involves only finitely many elements of S, and so there exists a finite subset $\{s_1, \ldots, s_n\}$ of S such that $k_i \in K(s_1, \ldots, s_n)$ for $0 \leqslant i \leqslant j$. Thus α is algebraic over $K(s_1, \ldots, s_n)$ and so $\{s_1, \ldots, s_n, \alpha\}$ is not algebraically independent over K, by Theorem 18.3. Consequently $S \cup \{\alpha\}$ is not algebraically independent over K, and S is maximal.

Just as every vector space has a basis, so does every extension $L:K$ have a transcendence basis. As in Theorem 2.1, we prove rather more.

Theorem 18.5 *Suppose that $L:K$ is an extension, that A is a subset of L such that $L:K(A)$ is algebraic and that C is a subset of A which is algebraically independent over K. Then there exists a transcendence basis B for L over K with $C \subseteq B \subseteq A$.*

Proof. The proof is very similar in nature to the proof of Theorem 2.1.

Indeed, if we replace the phrase 'linearly independent' by 'algebraically independent over K', we obtain a proof of the fact that there is a set B which is maximal among those which contain C, are contained in A and are algebraically independent over K.

The argument of Theorem 18.4 now shows that each element α of A is algebraic over $K(B)$, and so $K(A):K(B)$ is algebraic, by Corollary 2 to Theorem 4.6. As $L:K(A)$ is algebraic, $L:K(B)$ is algebraic (Theorem 4.7) and so B is a transcendence basis for L over K, by Theorem 18.4.

Consider the extension $\mathbb{R}:\mathbb{Q}$. If S is any countable subset of $\mathbb{Q}$, $\mathbb{Q}(S)$ is countable. If $\mathbb{R}:\mathbb{Q}(S)$ were algebraic, $\mathbb{R}$ would be countable (Exercise 4.7). Thus any transcendence basis for $\mathbb{R}$ over $\mathbb{Q}$ must be uncountable.

Note that, on the other hand, it follows from Theorem 18.5 that if $L:K$ is finitely generated over K then there must be a finite transcendence basis for L over K.

Exercise

18.5 Suppose that $L:K$ is an extension, and that L is finitely generated over K. Show that the field K_a of elements of L which are algebraic over K is finitely generated over K.

18.3   Transcendence degree

We now pursue further the parallelism with vector spaces. First we establish a version of the Steinitz exchange theorem (Theorem 1.3).

Theorem 18.6 *Suppose that $L:K$ is an extension, that $C=\{c_1,\ldots,c_r\}$ is a subset of L (with r distinct elements) which is algebraically independent over K and that $A=\{a_1,\ldots,a_s\}$ is a subset of L (with s distinct elements) such that $L:K(A)$ is algebraic. Then $r \leqslant s$, and there exists a set D, with $C \subseteq D \subseteq A \cup C$ such that $|D|=s$ and $L:K(D)$ is algebraic.*

Proof. We prove this by induction on r. The result is trivially true for $r=0$ (take $D=A$). Suppose that it is true for $r-1$. As the set $C_0=\{c_1,\ldots,c_{r-1}\}$ is algebraically independent over K, there exists a set D_0 with $C_0 \subseteq D_0 \subseteq A \cup C_0$ such that $|D_0|=s$ and $L:K(D_0)$ is algebraic. By relabelling A if necessary, we can suppose that

$$D_0=\{c_1,\ldots,c_{r-1},a_r,a_{r+1},\ldots,a_s\}.$$

As $L:K(D_0)$ is algebraic, c_r is algebraic over $K(D_0)$. As $\{c_1,\ldots,c_r\}$ is algebraically independent over K, c_r is transcendental over $K(c_1,\ldots,c_{r-1})$ (by Theorem 18.3). Thus $s \geqslant r$. Also, by Theorem 18.3 again,

$$E=\{c_1,\ldots,c_{r-1},c_r,a_r,a_{r+1},\ldots,a_s\}$$

is algebraically dependent over K. Using Theorem 18.3 once more, and using the fact that $\{c_1, \ldots, c_r\}$ is algebraically independent over K, we conclude that there exists t, with $r \leqslant t \leqslant s$, such that a_t is algebraic over $K(c_1, \ldots, c_r, a_r, \ldots, a_{t-1})$. Let $D = \{c_1, \ldots, c_r, a_r, \ldots, a_{t-1}, a_{t+1}, \ldots, a_s\}$. Then a_t is algebraic over $K(D)$, and so $K(E):K(D)$ is algebraic. As $E \supseteq D_0$, $L:K(E)$ is algebraic, and so $L:K(D)$ is algebraic, by Theorem 4.7. As $C \subseteq D \subseteq A \cup C$ and $|D| = s$, this completes the proof.

Corollary *If $L:K$ is an extension, and S and T are two transcendence bases for L over K then either S and T are both infinite or S and T have the same finite number of elements.*

If an extension $L:K$ has a finite transcendence basis, we define its *transcendence degree* to be the number of elements in the transcendence basis; otherwise we define the transcendence degree to be ∞.

18.4   The tower law for transcendence degree

Suppose that $M:L$ and $L:K$ are extensions. How is the transcendence degree of $M:K$ related to the transcendence degrees of $M:L$ and $L:K$?

Theorem 18.7 *Suppose that $M:L$ and $L:K$ are extensions, that A is a subset of L which is algebraically independent over K and that B is a subset of M which is algebraically independent over L. Then $A \cup B$ is algebraically independent over K.*

Proof. Let C be a finite subset of $A \cup B$. We can write

$$C = \{\alpha_1, \ldots, \alpha_r, \beta_1, \ldots, \beta_s\}$$

with $\alpha_i \in A$, $\beta_j \in B$. By Theorem 18.3, α_i is transcendental over $K(\alpha_1, \ldots, \alpha_{i-1})$ for $1 \leqslant i \leqslant r$ and β_j is transcendental over $L(\beta_1, \ldots, \beta_{j-1})$ for $1 \leqslant j \leqslant s$, and so β_j is transcendental over $K(\alpha_1, \ldots, \alpha_r, \beta_1, \ldots, \beta_{j-1})$ for $1 \leqslant j \leqslant s$. Thus C is algebraically independent over K, by Theorem 18.3. Since this holds for any finite subset of $A \cup B$, $A \cup B$ is algebraically independent over K.

Theorem 18.8 *Suppose that $M:L$ and $L:K$ are extensions, that A is a transcendence basis for L over K and that B is a transcendence basis for M over L. Then $A \cup B$ is a transcendence basis for M over K.*

Proof. By Theorems 18.4 and 18.7 it is enough to show that $M:K(A \cup B)$ is algebraic.

Since A is a transcendence basis for L over K, $L:K(A)$ is algebraic. Since $K(A) \subseteq K(A \cup B)$, it follows that $K(A \cup B)(L):K(A \cup B)$ is algebraic. As $K(A \cup B)(L) = L(B)$, this means that $L(B):K(A \cup B)$ is algebraic. But B is a transcendence basis for M over L, and so $M:L(B)$ is algebraic. Thus $M:K(A \cup B)$ is algebraic, by Theorem 4.7.

Corollary *If $M:L$ and $L:K$ are extensions, the transcendence degree of $M:K$ is the sum of the transcendence degrees of $M:L$ and $L:K$.*

18.5 Lüroth's theorem

Suppose that $L:K$ is a finitely generated extension which has transcendence degree r. If $\alpha_1, \ldots, \alpha_r$ is a transcendence basis for L over K, then $L:K(\alpha_1, \ldots, \alpha_r)$ is finite. If we can find a transcendence basis $\alpha_1, \ldots, \alpha_r$ for L over K such that $L = K(\alpha_1, \ldots, \alpha_r)$, then we say that L is *purely transcendental* over K. Even in particular cases, it is not easy to determine whether a finitely generated extension is purely transcendental or not (see Exercises 18.6 and 18.7 below). There is, however, one case where the problem can be solved in a straightforward way. The proof involves polynomials in two variables: first make sure that you are familiar with the contents of Section 3.7.

Theorem 18.9 (Lüroth's theorem) *Suppose that $K(t):K$ is a simple extension and that t is transcendental over K. If L is a subfield of $K(t)$ containing K then $L:K$ is a simple extension.*

Proof. We clearly need only consider the case where L is different from both K and $K(t)$. If $s \in L \setminus K$, we can write $s = p(t)/q(t)$ where p and q are non-zero polynomials in $K[x]$. Then $q(t)s - p(t) = 0$, and t is algebraic over L. Let m be the minimal polynomial of t over L. We can consider m as an element of $K(t)[x]$; by Theorem 3.11, there exists β in $K(t)$ such that $\beta m = f$, where

$$f = a_0(t) + a_1(t)x + \cdots + a_n(t)x^n$$

is a primitive polynomial in $K[t][x]$. Note that

$$n = \text{degree } m = [K(t):L].$$

Since m is monic, $\beta = a_n(t)$ and the terms $a_i(t)/a_n(t)$ are all in L; on the other hand, they are not all in K, since t is transcendental over K. There therefore exists i, with $0 \leqslant i < n$, such that $u = a_i(t)/a_n(t) \in L \setminus K$. We can write u as $g(t)/h(t)$ where g and h are relatively prime polynomials in $K[t]$.

Let $r = \max(\text{degree } g, \text{degree } h)$.

Then $[K(t):K(u)] = r$ (Exercise 5.5). As $K(u) \subseteq L$, this means that $r \geqslant n$. It also means that it is sufficient to show that $r \leqslant n$, for then it follows that $L = K(u)$.

We now consider the expression

$$l = g(t)h(x) - h(t)g(x).$$

As g and h are relatively prime, l is non-zero. Now $(h(t))^{-1}l \in L[x]$, and $(h(t))^{-1}l$ has t as a root: thus m divides $(h(t))^{-1}l$ in $L[x]$. This implies that f divides l in $K(t)[x]$. As f is primitive in $K[t][x]$, it follows from Corollary 1 to Theorem 3.13 that f divides l in $K[t][x]$. Thus there exists j in $K[t][x]$ such that $l = fj$.

We can consider f, l and j either as elements of $K[t][x]$ or as elements of $K[x][t]$: let us denote the degree in x by $\deg_x$ and the degree in t by $\deg_t$.

Now $\deg_t(l) \leqslant r$ and $\deg_t(f) \geqslant r$: since $f = lj$, $\deg_t(l) = \deg_t(f) = r$ and $\deg_t(j) = 0$. In other words, we can consider j as an element of $K[x]$. In particular, this means that j is primitive in $K[t][x]$, and so by Theorem 3.12 $l = fj$ is primitive in $K[t][x]$. As l is skew-symmetric in t and x, this implies that l is primitive in $K[x][t]$. But $j \in K[x]$, and j divides l; thus j must be a unit in $K[x]$, and so $j \in K$. Consequently

$$n = \deg_x(f) = \deg_x(l) = \deg_t(l) = \deg_t(f) \geqslant r,$$

and the theorem is proved.

Does Lüroth's theorem extend to purely transcendental extensions of higher transcendence degree? It can be shown that if t_1 and t_2 are algebraically independent over an algebraically closed field K, and M is a subfield of $K(t_1, t_2)$ for which $K(t_1, t_2):M$ is finite and separable, then M is purely transcendental over K. It can also be shown that a corresponding result does not hold for extensions of transcendence degree 3. These results involve polynomials in several variables in a more fundamental way than does Lüroth's theorem. The results really belong to algebraic geometry: they are discussed, for example, in the book by Hartshorne.[1]

Exercises

18.6 Suppose that $K(x, y):K$ is an extension with x transcendental over K and $x^2 + y^2 = 1$. Show that $K(x, y) = K(u)$, where $u = (1 + y)/x$.

18.7 Suppose that $n \geqslant 3$, that $K(x, y):K$ is an extension with x transcendental over K and $x^n + y^n = 1$ and that char K does not divide n. Suppose if possible that $K(x, y) = K(s)$.

(i) Show that there are relatively prime polynomials f, g and h in $K[x]$ such that max (degree f, degree g, degree h) $\geqslant 1$ and $f^n + g^n = h^n$.

(ii) Show that

$$f^{n-1} \big| (hDg - gDh) \quad \text{and} \quad g^{n-1} \big| (hDf - fDh),$$

and show (by considering degrees) that this is not possible.

[1] R. Hartshorne, *Algebraic Geometry*, Springer-Verlag, 1977.

19

Some further topics

In this chapter, we shall consider three further topics. Each has independent interest, and also shows how the theory that we have developed so far can be applied.

19.1 Generic polynomials

When we considered cubic polynomials, we saw that a variety of possibilities can arise. Some depend on the original field K: whether or not K contains cube roots of unity, for example. Others depend on special relationships between the coefficients: these can confuse the issue, and it is sensible to consider polynomials where this cannot happen.

Suppose that K is a field. Let $K(a_1, \ldots, a_n):K$ be an extension such that $\{a_1, \ldots, a_n\}$ is algebraically independent over K. Then the generic (monic) polynomial of degree n over K is the polynomial

$$x^n - a_1 x^{n-1} + \cdots + (-1)^n a_n.$$

Note that this is an element of $K(a_1, \ldots, a_n)[x]$, and *not* an element of $K[x]$. Note also that, since $\{a_1, \ldots, a_n\}$ is algebraically independent over K, there is no relationship between the coefficients: they are quite general.

We can also consider polynomials with general roots. Let $K(t_1, \ldots, t_n):K$ be another extension such that $\{t_1, \ldots, t_n\}$ is algebraically independent over K. Then we consider the polynomial

$$f = (x - t_1) \ldots (x - t_n).$$

Again, this is an element of $K(t_1, \ldots, t_n)[x]$, and *not* an element of $K[x]$. We can write

$$f = (x - t_1) \ldots (x - t_n) = x^n - s_1 x^{n-1} + \cdots + (-1)^n s_n,$$

where

$$s_1 = t_1 + \cdots + t_n,$$

$$s_2 = \sum_{1 \leqslant i < j \leqslant n} t_i t_j,$$

$$\vdots$$

$$s_n = t_1 t_2 \ldots t_n.$$

The expressions $s_1, \ldots, s_n$, considered as elements of $K[t_1, \ldots, t_n]$, are the *elementary symmetric polynomials in n variables*.

Suppose now that σ is a permutation of $\{1, \ldots, n\}$. Then σ determines an automorphism of $K(t_1, \ldots, t_n)$:

$$\text{if } \alpha = \frac{h(t_1, \ldots, t_n)}{g(t_1, \ldots, t_n)} \text{ then } \sigma(\alpha) = \frac{h(t_{\sigma(1)}, \ldots, t_{\sigma(n)})}{g(t_{\sigma(1)}, \ldots, t_{\sigma(n)})}$$

Let G be the group of all such automorphisms, and let L be the fixed field of G. Then, by Theorem 11.3, $K(t_1, \ldots, t_n):L$ is a Galois extension, with Galois group G.

Theorem 19.1 $L = K(s_1, \ldots, s_n)$.
Proof. We can consider f as an element of $K(s_1, \ldots, s_n)[x]$. Then $K(t_1, \ldots, t_n):K(s_1, \ldots, s_n)$ is a splitting field extension for f, and so $[K(t_1, \ldots, t_n):K(s_1, \ldots, s_n)] \leqslant n!$, by Theorem 7.3. But clearly $K(s_1, \ldots, s_n) \subseteq L$ and $[K(t_1, \ldots, t_n):L] = n!$: it therefore follows that $L = K(s_1, \ldots, s_n)$.

Corollary f *is irreducible over* $K(s_1, \ldots, s_n)$ *and* $\Gamma(f) \cong \Sigma_n$, *the group of permutations of* $\{1, \ldots, n\}$.

Theorem 19.2 *The elementary symmetric polynomials* $s_1, \ldots, s_n$ *are algebraically independent over* K.
Proof. The transcendence degree of $K(t_1, \ldots, t_n):K$ is n. As $K(t_1, \ldots, t_n):K(s_1, \ldots, s_n)$ is algebraic, $\{s_1, \ldots, s_n\}$ contains a transcendence basis for $K(t_1, \ldots, t_n)$ over K, by Theorem 18.5. This must have n elements, by the corollary to Theorem 18.6, and so it must be the whole of $\{s_1, \ldots, s_n\}$.

By Theorem 18.2, this means that there is an isomorphism of $K(a_1, \ldots, a_n)$ onto $K(s_1, \ldots, s_n)$, which sends a_i to s_i (for $1 \leqslant i \leqslant n$) and which sends the generic polynomial $x^n - a_1 x^n + \cdots + (-1)^n a_n$ to f. Thus f has the same properties as the generic polynomial. Summing up:

Theorem 19.3 *The generic polynomial*

$$x^n - a_1 x^{n-1} + \cdots + (-1)^n a_n$$

is irreducible over $K(a_1, \ldots, a_n)$. *It is separable, and its Galois group is isomorphic to* Σ_n. *It is solvable by radicals if and only if* $n \leqslant 4$.

We say that a polynomial f in $K[t_1, \ldots, t_n]$ is *symmetric* if $\sigma(f) = f$ for each σ in G. It follows from Theorem 19.1 that if f is a symmetric polynomial then f can be expressed as a *rational* expression in $s_1, \ldots, s_n$. This is clearly not a satisfactory result: let us improve on it.

Let us set $M_0 = K(s_1, \ldots, s_n)$ and for $1 \leqslant j \leqslant n$ let

$$M_j = M_0(t_1, \ldots, t_j) = M_{j-1}(t_j).$$

Thus $\quad K(t_1, \ldots, t_n) = M_n : M_{n-1} : \ldots : M_0 = K(s_1, \ldots, s_n)$.

Now let

$$f_j = \prod_{i=1}^{j} (x - t_i) \quad \text{and} \quad g_j = \prod_{i=j+1}^{n} (x - t_i).$$

Then $f_j \in K[t_1, \ldots, t_j][x]$ and $f_j g_j \in K[s_1, \ldots, s_n][x]$ so that, by Lemma 15.2,

$$g_j \in K[s_1, \ldots, s_n, t_1, \ldots, t_j][x] \subseteq M_j[x].$$

Now t_{j+1} is a root of g_j and so

$$[M_{j+1} : M_j] = [M_j(t_{j+1}) : M_j] \leqslant n - j.$$

But $[M_n : M_0] = [K(t_1, \ldots, t_n) : K(s_1, \ldots, s_n)] = n!$, and so it follows from the tower laws that $[M_{j+1} : M_j] = n - j$, for $0 \leqslant j < n$. In particular, this means that g_j is the minimal polynomial of t_{j+1} over M_j, and that

$$1, t_{j+1}, \ldots, t_{j+1}^{n-j-1}$$

is a basis for M_{j+1} over M_j.

We now show inductively that if $f \in K[s_1, \ldots, s_n, t_1, \ldots, t_j]$ then f can be written uniquely in the form

$$f = \sum p_{i_1 \ldots i_j} t_1^{i_1} \ldots t_j^{i_j}$$

where each $p_{i_1 \ldots i_j}$ is in $K[s_1, \ldots, s_n]$, and summation is over all multi-indices $i_1, \ldots, i_j$, where $0 \leqslant i_k \leqslant n - k$ for $1 \leqslant k \leqslant j$. The result is trivially true for $j = 0$. Suppose that it is true for j and that $f \in K[s_1, \ldots, s_n, t_1, \ldots, t_{j+1}]$. Since g_j is the minimal polynomial for t_{j+1} over M_j and $g_j \in K[s_1, \ldots, s_n, t_1, \ldots, t_j][x]$, we can write

$$t_{j+1}^{n-j} = a_0 + a_1 t_{j+1} + \cdots + a_{n-j-1} t_{j+1}^{n-j-1}$$

where the coefficients a_k are in $K[s_1, \ldots, s_n, t_1, \ldots, t_j]$. Substituting for t_{j+1}^{n-j} in f wherever it occurs, and repeating the procedure if necessary, it follows that we can write

$$f = b_0 + b_1 t_{j+1} + \cdots + b_{n-j-1} t_{j+1}^{n-j-1}$$

where the coefficients b_k are in $K[s_1, \ldots, s_n, t_1, \ldots, t_j]$; further as 1,

$t_{j+1}, \ldots, t_{j+1}^{n-j-1}$ is a basis for M_{j+1} over M_j, this expression is unique. The result now follows, by applying the inductive hypothesis to the coefficients b_k.

As an immediate consequence, we have

Theorem 19.4 *If f is a symmetric polynomial in $K[t_1, \ldots, t_n]$, there exists a unique g in $K[x_1, \ldots, x_n]$ such that*

$$f(t_1, \ldots, t_n) = g(s_1, \ldots, s_n).$$

This result can also be proved directly by elementary methods (Exercise 19.2). Note also that if f is a symmetric polynomial in $\mathbb{Z}[t_1, \ldots, t_n]$ then $g \in \mathbb{Z}[x_1, \ldots, x_n]$.

Corollary *Suppose that*

$$h = x^n - a_1 x^{n-1} + \cdots + (-1)^n a_n$$

is a monic polynomial in $K[x]$, with roots $\alpha_1, \ldots, \alpha_n$ in some splitting field extension. If f is a symmetric polynomial in $t_1, \ldots, t_n$ then

$$f(\alpha_1, \ldots, \alpha_n) = g(a_1, \ldots, a_n)$$

(where g is the polynomial of the theorem).

Exercises

19.1 Let G_j be the subgroup of G which fixes $t_1, \ldots, t_j$. Show that M_j is the fixed field of G_j.

19.2 Suppose that $f \in K[t_1, \ldots, t_n]$. Suppose that $a t_1^{k_1} \ldots t_n^{k_n}$ and $b t_1^{l_1} \ldots t_n^{l_n}$ are two terms in f (with $k_i \geqslant 0$, $l_i \geqslant 0$). There is a least integer j such that $k_j \neq l_j$. We say that $b t_1^{l_1} \ldots t_n^{l_n}$ *follows* $a t_1^{k_1} \ldots t_n^{k_n}$ if $l_j > k_j$, for this j. This defines a total order on the terms of f (the *lexicographic order*). The last term is called the *principal term* of f.
Suppose that f is symmetric and has principal term $a t_1^{k_1} \ldots t_n^{k_n}$.
(i) Show that $k_1 \geqslant k_2 \geqslant \cdots \geqslant k_n$.
(ii) Show that $a s_1^{k_1 - k_2} \ldots s_{n-1}^{k_{n-1} - k_n} s_n^{k_n}$ has the same principal term as f.

(iii) Show that f can be written as $F(s_1, \ldots, s_n)$, where $F \in K[x_1, \ldots, x_n]$.
(iv) Show that the expression in (iii) is unique.

19.2   The normal basis theorem

Suppose that $L:K$ is a Galois extension, with Galois group $G = \{\sigma_1, \ldots, \sigma_n\}$. We know that L is an n-dimensional vector space over K. The normal basis theorem asserts that we can find a basis for L over K in terms of $\sigma_1, \ldots, \sigma_n$.

We begin by establishing a result concerning *infinite* integral domains.

Theorem 19.5 *Suppose that R is an infinite subset of an integral domain S and that f is a non-zero element of $S[x_1, \ldots, x_n]$. Then there exists $(r_1, \ldots, r_n)$ in R^n such that $f(r_1, \ldots, r_n) \neq 0$.*

Proof. We prove this by induction on n. In the case where $n = 1$, f has only finitely many roots in F, the field of fractions of S, and so f has only finitely many roots in S: since R is infinite, there exists r in R such that $f(r) \neq 0$.

Suppose that the result is true for $n - 1$. $S[x_1]$ is an infinite integral domain, and we can consider f as a non-zero element of $S[x_1][x_2, \ldots, x_n]$. By the inductive hypothesis, there exist $r_2, \ldots, r_n$ in R such that $f(x_1, r_2, \ldots, r_n)$ is a non-zero element of $S[x_1]$; by the case where $n = 1$, there exists r_1 in R such that $f(r_1, \ldots, r_n) \neq 0$.

We now come to the normal basis theorem. Because we use Theorem 19.5 in the proof, we shall prove this only for infinite fields. The result is true for extensions $L:K$ where L and K are finite, but the proof, which exploits the fact that such an extension is cyclic, involves properties of endomorphisms of vector spaces which it is unreasonable either to expect the reader to know or to develop here.

Theorem 19.6 (The normal basis theorem) *Suppose that K is an infinite field, and that $L:K$ is a Galois extension, with Galois group $G = \{\sigma_1, \ldots, \sigma_n\}$. Then there exists l in L such that $(\sigma_1(l), \ldots, \sigma_n(l))$ is a basis for L over K.*

Proof. By relabelling G if necessary, we can suppose that σ_1 is the identity. Let us define $p(i, j)$ by the formula

$$\sigma_i \sigma_j = \sigma_{p(i,j)}$$

for $1 \leqslant i \leqslant n$, $1 \leqslant j \leqslant n$. Let $x_1, \ldots, x_n$ be indeterminates and let M be the $n \times n$ matrix $(x_{p(i,j)})_{i=1,j=1}^{n}$ with entries in $K[x_1, \ldots, x_n]$. Let $f = \det M$. Then $f \in K[x_1, \ldots, x_n]$, and f is non-zero, since x_1 occurs once in each row and once in each column, and so the coefficient of x_1^n is 1 or -1.

Now let $(b_1, \ldots, b_n)$ be a basis for L over K. By Theorem 11.2 the n trajectories

$$(T(b_j))_{j=1}^{n} = ((\sigma_i(b_j))_{i=1}^{n})_{j=1}^{n}$$

are linearly independent over L in L^n: in other words the $n \times n$ matrix $(\sigma_i(b_j))_{i=1,j=1}^{n}$ is invertible; let $C = (c_{ij})$ be its inverse.

We now set

$$g(x_1, \ldots, x_n) = f(\textstyle\sum_j \sigma_1(b_j)x_j, \ldots, \sum_j \sigma_n(b_j)x_j).$$

Since
$$f(x_1,\ldots,x_n)=g(\sum_j c_{1j}x_j,\ldots,\sum_j c_{nj}x_j),$$
g is a non-zero element of $L[x_1,\ldots,x_n]$. By Theorem 19.5 this means that there exist $k_1,\ldots,k_n$ in K such that $g(k_1,\ldots,k_n)\neq 0$. We set $l=k_1b_1+\cdots+k_nb_n$. Then

$$0\neq g(k_1,\ldots,k_n)=f(\sum_j \sigma_1(b_j)k_j,\ldots,\sum_j \sigma_n(b_j)k_j)$$
$$=f(\sigma_1(l),\ldots,\sigma_n(l))$$
$$=\det((\sigma_{p(i,j)}(l)))=\det((\sigma_i(\sigma_j(l)))).$$

This means that the matrix $(\sigma_i(\sigma_j(l)))$ is invertible, and so by Theorem 11.2 $(\sigma_1(l),\ldots,\sigma_n(l))$ is a basis for L over K.

Exercise

19.3 Show that the primitive nth roots of unity over $\mathbb{Q}$ form a normal basis for the splitting field of x^n-1 over $\mathbb{Q}$ if and only if n has no repeated prime factors.

19.3   Constructing regular polygons

We end this chapter by considering ruler-and-compass constructions again. We consider the following problem: given an integer $n>2$, can we construct a regular polygon with n sides of unit length, using ruler and compasses alone? Let us say that an integer n is *possible* if this can be done. Using the results of Chapter 6 (and its exercises) it should be clear that 3, 4, 5, 6, 8 and 10 are possible, while 7 and 9 are not. Note that if n is possible then so is $2n$, for we can certainly bisect angles.

Suppose that n is an integer greater than 2. We set $\theta_n=2\pi/n$,
$$x_n=\cos\theta_n,\quad y_n=\sin\theta_n,\quad \varepsilon_n=\cos\theta_n+i\sin\theta_n=e^{i\theta_n}.$$
The complex number ε_n is a primitive nth root of unity in $\mathbb{C}$. It is clear that (x_n,y_n) is constructible if and only if we can construct an angle θ_n, and it is equally clear that if we can construct angles θ and ϕ then we can construct angles $\theta+\phi$ and $|\theta-\phi|$. Suppose that m and n are possible and that m and n are relatively prime. Then there exist integers a and b such that $am+bn=1$; multiplying by $2\pi/mn$, it follows that $a\theta_n+b\theta_m=\theta_{mn}$, so that mn is possible. It is trivially true that if l is possible then so are all its divisors: in particular, if mn is possible, so are m and n. We therefore have the following:

Theorem 19.7 *Suppose that $n=2^t p_1^{a_1}\ldots p_r^{a_r}$, where $p_1,\ldots,p_r$ are distinct odd primes. Then n is possible if and only if p_i^b is possible for $1\leqslant b\leqslant a_i$, $1\leqslant i\leqslant r$.*

This result reduces the problem to determining when p^b is possible, when p is an odd prime.

Theorem 19.8 *Suppose that p is an odd prime. Then $n = p^b$ is possible if and only if $b = 1$ and p is of the form $2^k + 1$.*

Proof. First suppose that n is possible. Then (x_n, y_n) is constructible, and so $[\mathbb{Q}(x_n, y_n):\mathbb{Q}] = 2^r$ for some r, by Theorem 6.1. By the tower law,

$$[\mathbb{Q}(x_n, y_n, i):\mathbb{Q}] = [\mathbb{Q}(x_n, y_n, i): \mathbb{Q}(x_n, y_n)][\mathbb{Q}(x_n, y_n):\mathbb{Q}]$$
$$= 2^{r+1};$$

as $\varepsilon_n \in \mathbb{Q}(x_n, y_n, i)$, $[\mathbb{Q}(\varepsilon_n):\mathbb{Q}] = 2^s$ for some s.

Now, if $n = p^b$ is possible for some $b \geqslant 2$, then $m = p^2$ is possible, and so $[\mathbb{Q}(\varepsilon_m):\mathbb{Q}] = 2^s$, for some s. There is one primitive first root of unity, and $p - 1$ primitive pth roots of unity, and so there are $p^2 - p = p(p - 1)$ primitive mth roots of unity. Thus the cyclotomic polynomial Φ_m has degree $p(p - 1)$. But Φ_m is irreducible over $\mathbb{Q}$ (Theorem 15.3), and is the minimal polynomial for ε_m over $\mathbb{Q}$, so $[\mathbb{Q}(\varepsilon_m):\mathbb{Q}] = p(p - 1)$. This is not of the form 2^s: we conclude that $b = 1$ and that $n = p$. But Φ_p has degree $p - 1$, and so p is a prime of the form $2^s + 1$.

Conversely suppose that $n = 2^s + 1$ is prime. Then $[\mathbb{Q}(\varepsilon_n):\mathbb{Q}] = 2^s$, $\mathbb{Q}(\varepsilon_n):\mathbb{Q}$ is a splitting field extension for Φ_n, and $G = \Gamma(\mathbb{Q}(\varepsilon_n):\mathbb{Q})$ is cyclic of degree 2^s, by the corollary to Theorem 15.4. Let σ be a generator for G. Then, if $0 \leqslant t \leqslant s$, the group G_{s-t} generated by $\sigma^{2^{s-t}}$ has order 2^t, and so there are intermediate groups

$$\{e\} = G_s \subseteq G_{s-1} \subseteq \cdots \subseteq G_0 = G, \text{ with } |G_{j-1}/G_j| = 2, \text{ for } 1 \leqslant j \leqslant s.$$

Let L_j be the fixed field for G_j. Then

$$\mathbb{Q}(\varepsilon_n) = L_s : L_{s-1} : \ldots : L_0 = \mathbb{Q}$$

is a tower of fields, and $[L_j:L_{j-1}] = 2$ for $1 \leqslant j \leqslant s$.

We shall show that if $z \in \mathbb{Q}(\varepsilon_n)$ and $z = x + iy$ then (x, y) is constructible. We use induction on j. The result is true for elements of $L_0 = \mathbb{Q}$. Suppose that it holds for all elements of L_{j-1} and that $\alpha = \alpha_1 + i\alpha_2 \in L_j \backslash L_{j-1}$. Then the minimal polynomial m_α for α over L_{j-1} is a quadratic in $L_{j-1}[x]$:

$$m_\alpha = x^2 + 2bx + c.$$

Thus $\alpha = -b + v$, where $v^2 = \mu = b^2 - c \in L_{j-1}$. Let us set $b = b_1 + ib_2$, $v = v_1 + iv_2$ and $\mu = \mu_1 + i\mu_2 = re^{i\theta}$. By the inductive hypothesis, (μ_1, μ_2) is constructible: from this we can successively construct $(r, 0)$, $(r^{1/2}, 0)$ and $(r^{1/2} \cos(\theta/2), r^{1/2} \sin(\theta/2))$ (since we can bisect angles). As $(v_1, v_2) = \pm(r^{1/2} \cos(\theta/2), r^{1/2} \sin(\theta/2))$ and as (b_1, b_2) is constructible, by the inductive hypothesis, this means that (α_1, α_2) is constructible. This establishes the induction. This means that (x_n, y_n) is constructible (where $\varepsilon_n = x_n + iy_n$) and so n is possible.

Primes of the form $2^k + 1$ are known as *Fermat primes*. If $2^k + 1$ is a

Fermat prime, k must be of the form 2^l: for if $k = st$, with s odd and greater than 1,

$$2^k + 1 = 2^{st} + 1 = (2^t + 1)(2^{t(s-1)} - 2^{t(s-2)} + \cdots + 1).$$

The only known Fermat primes are 3, 5, 17, 257 and 65 537, which correspond to $k = 0, 1, 2, 3$ and 4. It has been shown, using computers, that if any other Fermat primes exist they must be larger than $10^{40\,000}$. To sum up:

Theorem 19.9 (Gauss) *It is possible to construct a regular polygon with n sides if and only if n is of the form $2^t p_1 \ldots p_r$, where $t \geqslant 0$ and $p_1, \ldots, p_r$ are distinct Fermat primes.*

20

The calculation of Galois groups

In Chapter 17, we saw that (provided that certain assumptions are made about the characteristic of the underlying field) a polynomial is solvable by radicals if and only if its Galois group is soluble. This result raises many questions. Given an irreducible separable polynomial in $K[x]$, can its Galois group be determined? Given an integer n, what are the possible Galois groups of an irreducible separable polynomial of degree n in $K[x]$? Given a finite group G and a field K, does there exist an irreducible separable polynomial in $K[x]$ whose Galois group is isomorphic to G? In particular, what are the answers when K is the field $\mathbb{Q}$ of rational numbers?

20.1 A procedure for determining the Galois group of a polynomial

Suppose that K is a field. Let $t_1, \ldots, t_n, x_1, \ldots, x_n$ be indeterminates. If $\sigma \in \Sigma_n$, let σ_t denote the permutation of $t_1, \ldots, t_n$ which sends t_i to $t_{\sigma(i)}$ for $1 \leqslant i \leqslant n$, and let σ_x denote the permutation of $x_1, \ldots, x_n$ which sends x_i to $x_{\sigma(i)}$ for $1 \leqslant i \leqslant n$. We first consider the polynomial

$$P = \prod_{\sigma \in \Sigma_n} (y - (t_1 x_{\sigma(1)} + \cdots + t_n x_{\sigma(n)}))$$

$$= \prod_{\sigma \in \Sigma_n} (y - (t_{\sigma(1)} x_1 + \cdots + t_{\sigma(n)} x_n)).$$

Grouping the terms in y together, we can write

$$P = \sum_{j=0}^{n!} c_j y^j,$$

with

$$c_j = \Sigma_m f_m x_1^{i_1} \ldots x_n^{i_n},$$

where each f_m is in $K[t_1, \ldots, t_n]$ and the summation is taken over all multi-indices $m = (i_1, \ldots, i_n)$ with $i_k \geqslant 0$ for $1 \leqslant k \leqslant n$ and $\sum_{k=1}^{n} i_k = n! - j$.

Now $\sigma_t(P) = P$, and so $\sigma_t(f_m) = f_m$, for each m and each σ in Σ_n. By Theorem 19.4 we can write

$$f_m = g_m(s_1, \ldots, s_n)$$

where s_j is the jth elementary symmetric polynomial in $t_1, \ldots, t_n$ and $g_m \in K[s_1, \ldots, s_n]$. Thus

$$P = \sum_{j=0}^{n!} (\Sigma_m g_m(s_1, \ldots, s_n) x_1^{i_1} \ldots x_n^{i_n}) y^j;$$

further, given n, it is possible to calculate the polynomials g_m explicitly.

Now suppose that

$$f = x^n - a_1 x^{n-1} + \cdots + (-1)^n a_n$$

is a monic polynomial in $K[x]$ which has distinct roots $\alpha_1, \ldots, \alpha_n$ in a splitting field extension L. We set $\beta = \alpha_1 x_1 + \cdots + \alpha_n x_n$. If $\sigma \in \Sigma_n$, we set

$$\sigma_x(\beta) = \alpha_1 x_{\sigma(1)} + \cdots + \alpha_n x_{\sigma(n)}$$

and $\sigma_\alpha(\beta) = \alpha_{\sigma(1)} x_1 + \cdots + \alpha_{\sigma(n)} x_n = (\sigma^{-1})_x(\beta)$. Note that, since f has distinct roots, $\sigma_x(\beta) \neq \tau_x(\beta)$ if $\sigma \neq \tau$. We consider the polynomial

$$F = \prod_{\sigma \in \Sigma_n} (y - \sigma_x(\beta)) = \prod_{\sigma \in \Sigma_n} (y - \sigma_\alpha(\beta)).$$

We obtain F by substituting α_j for t_j in P: thus

$$F = \sum_{j=0}^{n!} (\Sigma_m g_m(a_1, \ldots, a_n) x_1^{i_1} \ldots x_n^{i_n}) y^j$$

and so $F \in K[y, x_1, \ldots, x_n]$.

As $K[y, x_1, \ldots, x_n]$ is a unique factorization domain (Corollary 2 to Theorem 3.13), we can write $F = F_1 \ldots F_k$, where each F_j is irreducible in $K[y, x_1, \ldots, x_n]$. Considering the F_j as elements of $L[y, x_1, \ldots, x_n]$, we can write each F_j as

$$F_j = \prod_{\sigma \in A_j} (y - \sigma_x(\beta))$$

where $A_1, \ldots, A_k$ is a partition of Σ_n. By labelling the F_j appropriately, we can suppose that the identity permutation is in A_1: thus $y - \beta$ divides F_1 in $L[y, x_1, \ldots, x_n]$.

Now suppose that $\sigma \in \Sigma_n$. Then

$$\sigma_x F = (\sigma_x F_1)(\sigma_x F_2) \ldots (\sigma_x F_k)$$

But $\sigma_x F = F$. Thus σ induces a permutation of the irreducible factors $F_1, \ldots, F_k$. Let

$$G = \{\sigma : \sigma_x F_1 = F_1\};$$

it is clear that G is a subgroup of Σ_n.

Theorem 20.1 *The group G is isomorphic to the Galois group $\Gamma_K(f)$.*
Proof. First note that

$$A_1 = \{\sigma : y - \sigma_x \beta \text{ divides } F_1 \text{ in } L[y, x_1, \ldots, x_n]\}$$
$$= \{\sigma : y - \beta \text{ divides } (\sigma^{-1})_x F_1 \text{ in } L[y, x_1, \ldots, x_n]\}$$
$$\{\sigma : (\sigma^{-1})_x F_1 = F_1\} = G.$$

Now let

$$H = \prod_{\sigma \in \Gamma_K(f)} (y - \sigma_\alpha(\beta)) = \prod_{\sigma \in \Gamma_K(f)} (y - \sigma_x(\beta)).$$

If $\tau \in \Gamma_K(f)$, $\tau H = H$, so $H \in K[y, x_1, \ldots, x_n]$. H divides F in $L[y, x_1, \ldots, x_n]$ and so H divides F in $L(x_1, \ldots, x_n)[y]$. Thus H divides F in $K(x_1, \ldots, x_n)[y]$ (Lemma 15.2) and so H divides F in $K[y, x_1, \ldots, x_n]$ (Corollary 1 of Theorem 3.13). This means that we can write H as a product of certain of the irreducible factors $F_1, \ldots, F_k$ of F. As $y - \beta$ divides H in $L[y, x_1, \ldots, x_n]$, F_1 must be one of these factors: thus F_1 divides H in $K[y, x_1, \ldots, x_n]$, and so $G \subseteq \Gamma_K(f)$.

Conversely, if $\tau \in \Gamma_K(f)$

$$\tau_x(F_1) = \prod_{\sigma \in A_1} (y - \tau_x \sigma_x(\beta))$$
$$= \prod_{\sigma \in A_1} (y - (\tau^{-1})_\alpha \sigma_x(\beta))$$
$$= (\tau^{-1})_\alpha \prod_{\sigma \in A_1} (y - \sigma_x(\beta)) = (\tau^{-1})_\alpha(F_1).$$

But $F_1 \in K[y, x_1, \ldots, x_n]$, and so $(\tau^{-1})_\alpha(F_1) = F_1$. Thus $\tau \in G$ and so $\Gamma_K(f) \subseteq G$.

This completes the proof. Notice that if F_i is another irreducible factor of F, there exists τ in Σ_n such that $\tau(F_1) = F_i$, and

$$\tau G \tau^{-1} = \{\sigma : \sigma_x(F_i) = F_i\};$$

thus each of the groups $\{\sigma : \sigma_x(F_i) = F_i\}$ is isomorphic to $\Gamma_K(f)$.

Suppose now that g is a polynomial in $\mathbb{Z}[x]$. By Theorem 5.1, there is an algorithm for expressing g as a product of irreducible factors

$$g = g_1^{r_1} \ldots g_j^{r_j}.$$

Let $f = g_1 \ldots g_j$. Then $\Gamma_\mathbb{Q}(f) = \Gamma_\mathbb{Q}(g)$, and f has distinct roots in a splitting field. We can calculate the polynomial F. $F \in \mathbb{Z}[y, x_1, \ldots, x_n]$. Now it is a straightforward matter to extend the argument of Theorem 5.1 to show that an algorithm exists to express F as a product of irreducible factors in $\mathbb{Z}[y, x_1, \ldots, x_n]$. Having found an irreducible factor F_1, it remains to work through the elements of Σ_n to determine which of them fix F_1. Thus an algorithm exists for calculating the Galois group of an element of $\mathbb{Z}[x]$.

This result is theoretically important, but the algorithm is much too complicated to be of any practical use. As we shall see when we consider the quintic, though, it does suggest ways of deciding between various possibilities. It also leads on to the next result, which is of great practical value.

Theorem 20.2 *Suppose that f is a monic polynomial in $\mathbb{Z}[x]$ and that p is a prime. Let $\bar{f}$ be the corresponding element of $\mathbb{Z}_p[x]$. If f has distinct roots in a splitting field extension L, then the cyclic group $\Gamma_{\mathbb{Z}_p}(\bar{f})$ is isomorphic to a subgroup of $\Gamma_{\mathbb{Q}}(f)$.*

Proof. Suppose that

$$f = x^n - a_1 x^{n-1} + \cdots + (-1)^n a_n$$

Let $\alpha_1, \ldots, \alpha_n$ be roots of f in a splitting field extension $L:\mathbb{Q}$, let $\beta = \alpha_1 x_1 + \cdots + \alpha_n x_n$ and let

$$F = \prod_{\sigma \in \Sigma_n} (y - \sigma_x \beta).$$

Similarly let $\gamma_1, \ldots, \gamma_n$ be roots of $\bar{f}$ in a splitting field extension $M:\mathbb{Z}_p$ and let $\delta = \gamma_1 x_1 + \cdots + \gamma_n x_n$. As

$$F = \sum_{j=0}^{n!} (\Sigma_m g_m(a_1, \ldots, a_n) x_1^{i_1} \ldots x_n^{i_n}) y^j,$$

$F \in \mathbb{Z}[y, x_1, \ldots, x_n]$ and

$$\bar{F} = \prod_{\sigma \in \Sigma_n} (y - \sigma_x \delta).$$

Let

$$F_1 = \prod_{\sigma \in \Gamma_{\mathbb{Q}}(f)} (y - \sigma_x \beta), \qquad G_1 = \prod_{\sigma \in \Gamma_{\mathbb{Z}_p}(\bar{f})} (y - \sigma_x \delta).$$

By Theorem 20.1, F_1 is an irreducible factor of F in $\mathbb{Q}[y, x_1, \ldots, x_n]$ and G_1 is an irreducible factor of $\bar{F}$ in $\mathbb{Z}_p[y, x_1, \ldots, x_n]$. $\bar{F}_1$ is a (not necessarily irreducible) factor of $\bar{F}$ in $\mathbb{Z}_p[y, x_1, \ldots, x_n]$ and $y - \delta$ divides both G_1 and $\bar{F}_1$ in $M[y, x_1, \ldots, x_n]$, and so G_1 divides

$$\bar{F}_1 = \prod_{\sigma \in \Gamma_{\mathbb{Q}}(f)} (y - \sigma_x \delta)$$

in $\mathbb{Z}_p[y, x_1, \ldots, x_n]$. This means that $\Gamma_{\mathbb{Z}_p}(\bar{f}) \subseteq \Gamma_{\mathbb{Q}}(f)$.

20.2   The soluble transitive subgroups of Σ_p

Suppose that f is an irreducible polynomial of prime degree p in $K[x]$ (where char $K \neq p$). The Galois group $\Gamma_K(f)$ is isomorphic to a transitive subgroup of Σ_p. f is solvable by radicals if and only if $\Gamma_K(f)$ is soluble. It is therefore desirable to know what the soluble transitive subgroups of Σ_p are.

We can consider Σ_p as acting on a set S with p elements. Suppose that G is a transitive subgroup of Σ_p and that H is a normal subgroup of G other than $\{e\}$. We shall show that H is also a transitive subgroup of Σ_p. If $x \in S$, let

$$O_H(x) = \{\sigma x : \sigma \in H\}$$

be the *orbit* of x under H. The relation '$x \sim x'$ if there exists σ in H such that $\sigma x = x''$ is an equivalence relation, and $O_H(x)$ is the equivalence class of x under this relation. Thus any two orbits are either identical or disjoint. If x and y are in S, then since G is transitive there exists τ in G such that $\tau x = y$. If $x' = \sigma x \in O_H(x)$ then

$$\tau x' = \tau \sigma x = \tau \sigma \tau^{-1} y$$

so that, since $\tau \sigma \tau^{-1} \in H$, $\tau x' \in O_H(y)$. Thus τ is a one–one mapping of $O_H(x)$ into $O_H(y)$. Similarly τ^{-1} is a one–one mapping of $O_H(y)$ into $O_H(x)$, and so $O_H(x)$ and $O_H(y)$ have the same number of elements. Not every orbit is a one-point set, since $H \neq \{e\}$. Since p is a prime, it follows that $O_H(x) = S$ for each x in S: in other words, H is a transitive subgroup of Σ_p.

Suppose now that G is a soluble transitive subgroup of Σ_p, and that

$$\{e\} = G_n \subseteq G_{n-1} \subseteq \cdots \subseteq G_0 = G$$

is a finite series of subgroups such that $G_i \lhd G_{i-1}$ for $1 \leqslant i \leqslant n$, G_{i-1}/G_i is cyclic for $1 \leqslant i \leqslant n$ and $G_{n-1} \neq \{e\}$. Repeated application of the above result shows that G_{n-1} is a cyclic transitive subgroup of Σ_p. G_{n-1} is therefore cyclic of order p. Let σ be a generator of G_{n-1}. We can write S as

$$S = \{0, 1, 2, \ldots, p-1\}$$

in such a way that $\sigma(j) = j + 1 \pmod{p}$. It will now be convenient to identify S with the finite field $\mathbb{Z}_p$.

It is easy to verify that the set of *affine* transformations of $\mathbb{Z}_p$ of the form

$$\tau_{(a,b)}(k) = ak + b,$$

where $a \in \mathbb{Z}_p^*$ and $b \in \mathbb{Z}_p$, forms a subgroup W of Σ_p of order $p(p-1)$. The mapping $\tau_{(a,b)} \to a$ is a homomorphism of W onto the multiplicative group $\mathbb{Z}_p^*$ with kernel $G_{n-1} = \{\tau_{(1,b)} : b \in \mathbb{Z}_p\}$. Thus G_{n-1} is a normal subgroup of W; the group G_{n-1} is cyclic of order p and W/G_{n-1} is cyclic of order $p-1$, so that W is soluble.

We shall show that $G \subseteq W$. Suppose that $G_j \subseteq W$, and that $\tau \in G_{j-1}$. Then $\tau \sigma \tau^{-1} \in G_j$ (where, as before, $\sigma(k) = k + 1 \pmod{p}$) and so $\tau \sigma \tau^{-1} = \tau_{(a,b)}$ for some a and b. Now $\tau \sigma \tau^{-1}$ has order p, and so it permutes the p elements of $\mathbb{Z}_p$ cyclically: thus the equation

$$\tau \sigma \tau^{-1}(x) = ax + b = x$$

has no solution in $\mathbb{Z}_p$; this happens if and only if $a = 1$ and $b \neq 0$. In other words $\tau\sigma\tau^{-1}$ is not the identity element of G_{n-1}. Thus if $k \in \mathbb{Z}_p$,

$$\tau(k+1) = \tau\sigma(k) = \tau\sigma\tau^{-1}\tau(k) = \tau(k) + b,$$

and so $\tau(k) = bk + \tau(0)$. Thus $\tau \in W$, and $G_{j-1} \subseteq W$. Thus it follows by induction that $G = G_0 \subseteq W$.

To sum up:

Theorem 20.3 *Suppose that G is a soluble transitive subgroup of Σ_p (where p is a prime). G contains a normal transitive cyclic subgroup T of order p, and G/T is cyclic, with order dividing $p-1$. Further, there exists a subgroup F of the cyclic group $\mathbb{Z}_p^*$ such that G is isomorphic to the group of affine transformations $\{\tau_{(a,b)} : a \in F, b \in \mathbb{Z}_p\}$ of $\mathbb{Z}_p$, where $\tau_{(a,b)}(k) = ak + b$.*

Recall that in Section 7.3 we showed that the degree of a splitting field extension for $x^p - 2$ over $\mathbb{Q}$ is $p(p-1)$, and so the Galois group of $x^p - 2$ is isomorphic to the group W. It is remarkable that a polynomial of such a simple form has a Galois group which is as large as possible.

Exercises

20.1 The polynomial $f = x^7 + 9x^2 + 7$ is irreducible over $\mathbb{Q}$ (this can be checked, using Theorem 5.1). Let $\bar{f}$ be the corresponding polynomial in $\mathbb{Z}_7[x]$. Show that $\bar{f}$ splits over $\mathbb{Z}_7$ as the product of an irreducible quartic and three linear factors (use Exercise 15.6) and show that f is not solvable by radicals.

20.2 Show that if G is a soluble transitive subgroup of Σ_p (where p is a prime) then every element of G other than the identity fixes at most one point.

20.3 Show that if f is an irreducible polynomial in $\mathbb{Q}[x]$ of odd prime degree p which is solvable by radicals then either all the roots of f are real or f has exactly one real root. Show that if $p = 4k + 3$ then the discriminant can be used to distinguish the two possibilities. What happens if $p = 4k + 1$?

20.4 This question needs the following result from group theory. If G is a group of order pq, where p is a prime which does not divide q then G has a subgroup of order p.

Let G be a transitive subgroup of Σ_p (where p is a prime).

(i) Show that $|G| = pq$, where p does not divide q. (Hint: Consider the subgroup H of G which fixes a certain point and consider the index of H in G.)

(ii) Show that if G has at least two subgroups of order p then there exists an element of G, other than the identity, which fixes two points and so G is not soluble.

(iii) Show that if G has exactly one subgroup K of order p, then $K \lhd G$ and G is soluble.

20.5 Suppose that f is an irreducible polynomial of prime degree p in $K[x]$, and that char $K \neq p$. Let $L:K$ be a splitting field extension for f. Show that f is solvable by radicals if and only if whenever α and β are distinct roots of f then $L = K(\alpha, \beta)$.

20.3 The Galois group of a quintic

Let us now consider the possible Galois groups of irreducible quintics. Suppose that f is an irreducible separable quintic in $K[x]$. If f is solvable by radicals then, by the results above, $\Gamma_K(f)$ is isomorphic either to W, which has order 20, or to D_{10}, the group of rotations and reflections of a regular pentagon, which has order 10, or to the cyclic group $\mathbb{Z}_5$ of order 5. If f is not solvable by radicals then $\Gamma_K(f)$ is isomorphic to the alternating group A_5 or the full symmetric group Σ_5.

Let us list some examples of irreducible quintic polynomials in $\mathbb{Z}[x]$, together with their Galois groups, to show that all possibilities can occur:

(a) $x^5 + x^4 - 4x^3 - 3x^2 + 3x + 1$ $\mathbb{Z}_5$
(b) $x^5 - 5x + 12$ D_{10}
(c) $x^5 - 2$ W
(d) $x^5 + 20x + 16$ A_5
(e) $x^5 - 4x + 2$ Σ_5

We have already discussed examples (c) and (e). Example (a) is obtained by considering a primitive 11th root of unity, α say. α has minimal polynomial $\sum_{n=0}^{10} x^n$, which has cyclic Galois group $\mathbb{Z}_{10}$. This has a subgroup of order 2 and the fixed field of this has Galois group $\mathbb{Z}_5$. The fixed field is generated by $\alpha + \alpha^{-1}$, and example (a) is the minimal polynomial of this.

We now turn to example (d). This has discriminant $2^{16} \, 5^6$ (use Exercise 14.3), so that its Galois group is contained in A_5. In $\mathbb{Z}_7$ the corresponding polynomial is

$$(x + 2)(x + 3)(x^3 + 2x^2 - 2x - 2)$$

and the cubic is irreducible. It follows from Theorem 20.2 that the Galois group of example (d) contains a cyclic subgroup of order 3, and so it must be A_5.

Finally we consider example (b). This has discriminant $2^{12} 5^6$, and in $\mathbb{Z}_3$ the corresponding polynomial factorizes as

$$x(x^2 + x - 1)(x^2 - x - 1)$$

and so has Galois group $\mathbb{Z}_2$. Using Theorem 20.2 again, we see that this means that the polynomial f of example (b) has Galois group D_{10} or A_5. How can one distinguish between these possibilities? Use of Theorem 20.1 would involve considering polynomials of degree 120, which is clearly impracticable. One way to proceed is the following. Let $\alpha_1, \ldots, \alpha_5$ be the roots of the polynomial in a splitting field L. We consider the ten elements $\alpha_i + \alpha_j$ of L, with $1 \leqslant i < j \leqslant 5$. It is easy to verify that these are distinct, so that the polynomial

$$g = \prod_{1 \leqslant i < j \leqslant 5} (x - (\alpha_i + \alpha_j))$$

has ten distinct roots. g is invariant under $\Gamma_\mathbb{Q}(f)$, and so $g \in \mathbb{Q}[x]$. It is clear that g splits over L. Since

$$\alpha_1 = \left(\sum_{i=1}^{5} \alpha_i \right) - (\alpha_2 + \alpha_3) - (\alpha_4 + \alpha_5)$$

and since similar equations are satisfied by $\alpha_2, \alpha_3, \alpha_4$ and α_5 it follows that $L:\mathbb{Q}$ is a splitting field extension for g. Suppose that f had Galois group A_5. Then the Galois group would act transitively on the roots of g, and so g would be irreducible. Using a computer it can be shown that this is not so.

I am grateful to Leonard Soicher for showing me examples (b) and (d).

Exercise

20.6 Suppose that f is a quintic in $\mathbb{Q}[x]$ whose Galois group contains D_{10}. Show that the ten elements $\alpha_i + \alpha_j$ $(1 \leqslant i < j \leqslant 5)$ are distinct (where $\alpha_1, \ldots, \alpha_5$ are the roots of f in $\mathbb{C}$).

20.4   Concluding remarks

The examples of the previous section show that there are irreducible quintics in $\mathbb{Q}[x]$ with all possible Galois groups. This raises the question: given a group G, does there exist an irreducible polynomial in $\mathbb{Q}[x]$ which has G as its Galois group? This is an extremely difficult problem which has not yet been solved. In 1954, Safarevich showed that the answer is 'Yes' when G is a soluble group.

Galois theory has a long and distinguished history: nevertheless, many interesting problems remain.

INDEX